John Joseph Griffin

Scientific Handicraft

A descriptive, illustrated, and priced catalogue of apparatus, suitable for the performance of elementary experiments in physics

John Joseph Griffin

Scientific Handicraft
A descriptive, illustrated, and priced catalogue of apparatus, suitable for the performance of elementary experiments in physics

ISBN/EAN: 9783337377281

Printed in Europe, USA, Canada, Australia, Japan

Cover: Foto ©berggeist007 / pixelio.de

More available books at **www.hansebooks.com**

SCIENTIFIC HANDICRAFT:

A Descriptive, Illustrated, and Priced

CATALOGUE OF APPARATUS,

SUITABLE FOR THE PERFORMANCE

OF

ELEMENTARY EXPERIMENTS IN PHYSICS.

BY

JOHN JOSEPH GRIFFIN, F.C.S.,

AUTHOR OF "CHEMICAL RECREATIONS;" "CHEMICAL HANDICRAFT;" ETC.

VOLUME FIRST:
MECHANICS, HYDROSTATICS, HYDRODYNAMICS,
AND
PNEUMATICS.

LONDON:
PUBLISHED BY JOHN J. GRIFFIN AND SONS,
CHEMICAL & PHILOSOPHICAL INSTRUMENT MAKERS,
22, GARRICK STREET, W.C.
1873.

PREFACE.

The nature of this Work is fully explained in the title-page. It is "a descriptive, illustrated, and priced CATALOGUE of APPARATUS suitable for the performance of *Elementary Experiments* in *Physics*." The present volume contains the subjects of Mechanics, Hydrostatics, Hydrodynamics, and Pneumatics. If it meets with public approval, companion volumes will be published on Acoustics, Heat, Light, Electricity, Magnetism, and Galvanism. These works will be strictly practical. They will avoid theoretical disquisitions, and will relate to Apparatus and Experiments only, and especially to apparatus of a cheap character, and to experiments suitable for elementary instruction.

Nearly all the Instruments and Experiments described in this volume have been made in the workshops of the Publishers, and tried in the presence of the Editor; and the figures and descriptions have been drawn to agree with the Instruments that had been submitted to these special trials.

JOHN J. GRIFFIN.

22, GARRICK STREET, W.C.
   London, July, 1873.

ADVERTISEMENT.

The Instruments described in this Work are offered for sale at the prices affixed to each. These prices are nett, for ready money. The expense of packing-cases and packing-materials is charged to purchasers. We employ very careful and experienced persons to pack the instruments securely; but we do not hold ourselves responsible for any breakage that may take place during the carriage of the goods from our manufactory to their places of destination. The packing-cases and packing-materials are charged at the lowest possible price, and we are compelled to intimate that we cannot take them back, nor allow their value to be deducted from our bills.

Orders from strangers in the country must be accompanied by a REMITTANCE, or by a reference to some person in London. Post-office orders are to be made payable at CHARING CROSS. Sums *under* 5s. may be remitted in postage stamps.

Foreign orders must either be accompanied by a remittance, or by instructions for payment in London, on delivery of the bills of lading, without which payment the orders cannot be executed. A large stock of the apparatus described in this work, both British and Foreign, being kept ready for immediate delivery, shipping orders for comprehensive collections can be executed with promptitude.

Such articles in this catalogue as are subject to frequent variations in price are at all times charged according to the lowest market value. The present prices supersede all those in catalogues of earlier date.

JOHN J. GRIFFIN AND SONS.

22, GARRICK STREET, W.C.
London, July, 1873.

CONTENTS.

MECHANICS.

	PAGE
THE MECHANICAL POWERS	1
Auxiliary apparatus	1
suspension board or black rail	1
schoolroom blackboard	1
pair of iron hooks for suspending apparatus	1
light brass scale pans	2
weights $\frac{1}{10}$ lb. each, two sorts	2
iron pins and hooks, for pulleys	2
cord suitable for pulleys	2
THE LEVER	2
lever with a wedge-shaped fulcrum	2
extemporised balance	3
lever for suspension	3
varieties of levers	4
lever of the first kind	4
lever of the second kind	5
lever of the third kind	6
Gravesende's suspended lever	6
experiments with it	6
THE PULLEY	7
pulley sheave, block, and axle	7
the pulley frame	7
sets of pulleys mounted	7
restrictions on the action of pulleys	7
simple pulleys, a set of four	7
experiments with simple pulleys	7
moveable pulleys and fixed pulleys	7
systems of pulleys	8
pair of long three-sheave pulleys	8
pair of square three-sheave pulleys	10
White's concentric pulley	10
pulley frame, and set of brass pulleys	10

	PAGE
THE INCLINED PLANE	10
THE WEDGE	11
THE WHEEL AND AXLE	11
Compound Wheel and Axle	12
THE SCREW	13
Small set of models of mechanical powers	13
APPARATUS TO ILLUSTRATE THE CENTRE OF GRAVITY	14
Centre of gravity of two parallelopipeds	14
of a leaning tower	14
of an irregular board	14
Centre of gravity not necessarily within the volume of a solid body	14
The equilibriste	15
The bowl-about	15
Double cone to run up an inclined plane	15
DEGREES OF STABILITY	16
Cone for showing states of equilibrium	16
Inertia apparatus	16
Tin pan and dried peas	16
Plumb-line	16
The pendulum	16
Experiments with the pendulum	17
The parallelogram of forces	18
Adhesion	19
Adhesion plates	19
Toys illustrative of adhesion	19
Percussion	19
Percussion machines	19
LAWS OF FALLING BODIES	20
Attwood's Fall Machine	20
WHIRLING TABLE OR CENTRIFUGAL MACHINE	22
Adjuncts for twelve experiments with the whirling table to	

	PAGE
illustrate the laws of central forces	23
Atmospheric optic marvel	26
The gyroscope or rotating apparatus	27
Gyroscope with additions for various experiments	27
Gyroscopic balance	28
Gyroscopic top	28
Vibrating wire	28
Model of a centrifugal railway	29
Constructive mechanics, models of four joints	29
Parallel motion	29
Intermittent motion	29
A train of wheels	29
Hooke's universal joint	30
Forge or tilt hammer	30
Lock and key, a model	31
BRITTLENESS illustrated	31
Prince Rupert's drops	31
Bolognian flask	31
Dialysis	31
Graham's dialyser	31
Graham's osmometer	31
Cathetometer	31
Vernier	32
Clinometer	32
BALANCES	32
large hydrostatic balance	32
balance in glass case for accurate weighings	33
balance for weighing small quantities up to 8 ounces in mahogany box	34
Commercial balances	34

	PAGE
Balance for weighing small quantities up to ½ ounce	34
for common weighings up to 2 lb.	34
WEIGHTS.	
Accurate decimal grain weights from 600 grains to ·10 grain, in a mahogany box	34
Grain weights, less accurate than the above, the series from 600 grains to ½ grain, in a box	34
Pound pile of avoirdupois, from 1 lb. to ½ oz.	34
ditto, in cast-iron	34
Single cast-iron weights from 1 lb. to 28 lb.	35
Accurate centigrade weights	35
Gramme weights from 50 grammes to ·001 gramme	35
Gramme weights, less accurate than the above, for ordinary use, from 200 grammes to ·01 gramme, in a box	35
Gramme weights, in cast-iron, in a pile, from 1000 grammes to ½ gramme	35
Standard gramme weights, in brass cubes	35
Graduated liquid measures	35
ounce measures	35
gramme measures	35
CHROMO-LITHOGRAPHIC LINEAR SCALES	36
inch scale to 40 inches	36
centimetre scale to 1 metre	36

HYDROSTATICS.

Water rises to a level in communicating vessels	37
apparatus, figures 201, 202, 203, 204	37
apparatus, figure 205	38
apparatus, figure 206	38
apparatus, figure 207	39
Spirit level, unmounted	39
Spirit level, mounted in mahogany	39
Level of two different liquors in communicating vessels	39
apparatus, fig. 213, jar with two tubes	40
apparatus, fig. 214, graduated pouret	40
apparatus, fig. 215, U tube graduated	40
apparatus, fig. 216, ditto unequal limbs	41
communicating vessels for three liquids	42
determination of the specific gravity of oil of vitriol and of ether	42

	PAGE
Boyle's inverted U tube for estimating the specific gravities of liquids	43
phial of the four elements	43
Pressure of water in all directions	44
apparatus, figure 221, glass globe with four tubes	44
pressure apparatus of metal with five tubes	45
glass pressure apparatus with four tubes	47
incompressibility of water	46
why water rises in pumps	46
Lateral pressure of liquids	47
jar for showing lateral pressure	47
Barker's mill	47
tourniquet hydraulique	48
Upward pressure of water	48
fig. 231, cylinder and disk	49
„ 234, cylinder and bladder	49
„ 235, hydrostatic bellows	49
„ 236, 237, ditto	50
Bramah's hydrostatic press	50
fig. 241, glass model	51
„ 242, small brass model	51
„ 243, large brass model	52
Hydrostatic paradox	52
Pascal's hydrostatic principle	52
Pascal's apparatus as modified by Masson	53
Haldat's pressure apparatus	54
Apparatus for weighing a cylinder, or a cone of water, in a vessel without a fixed bottom	55
experiments with a glass cone, 10 inches long	56
determination of the capacity of the cone in ounces of water	56
weight of the column of water that fills the cone when it is weighed with the narrow end downwards	56
weight of the column of water that fills the cone when it is weighed with the wide end downwards	57
paradoxical results	57
Specific gravity	57

	PAGE
equilibrium of solids when immersed in liquid	57
principle of Archimedes, viz. that every solid when immersed in a liquid loses a portion of its weight equal to the weight of the liquid which it displaces	57
exhibition of differences in specific gravities	58
three liquids of different densities	58
three solids of different densities	58
experimental demonstrations of the principle of Archimedes	59
Exhibition of various methods of estimating the specific gravity both of solids and liquids	61
estimation of the specific gravity of a solid body by means of the hydrostatic balance	61
estimation of the specific gravity of a liquid by means of the hydrostatic balance	62
estimation of the specific gravity of a solid body by weighing it in a bottle	62
estimation of the specific gravity of a liquid by weighing it in a bottle	63
Various forms of specific gravity bottles for liquids	65
prices of specific gravity bottles	65
practical remarks on the choice and use of specific gravity bottles	67
estimation of the specific gravity of a liquid, by weighing in a beaker a quantity first measured by a pipette	69
Mohr's hydrostatic balance for the rapid estimation of the specific gravity of liquids	69
Mohr's method of estimating the specific gravity of a solid by measuring the water it displaces	71
Nicholson's hydrometer for taking the specific gravities of minerals and other solids	72

	PAGE
THE HYDROMETER	73
Estimation of the specific gravity of liquids by means of the hydrometer	73
various forms of the glass hydrometer	74
trial jars for the liquors to be tested	74
detailed account of the treatment of percentage solutions, especially those of alcohol	75
results of trials by the bottle	75
necessity for rapid testing in business	76
characteristics of the different forms of hydrometers	76
and of trial jars	76
varieties of hydrometer scales	77
specific gravity scales	77
Twaddell's scale	77
Baumé's scale	77
chemical percentage scales	77
scales adapted for use in hot climates	77
rules to be observed in using hydrometers	78
Price list of a few hydrometers	78
FLOTATION	79
estimation of the weight of floating bodies by the measurement of the water they displace	79
equilibrium of a floating body	80

HYDRODYNAMICS.

	PAGE
CAPILLARITY	81
Apparatus for experiments on capillary action	82
Capillary tubes	82
Capillary plates	82
PRESSURE OF LIQUIDS PROPORTIONATE TO DEPTH	83
Spouting jars for Torricelli's experiment	83
THE PIPETTE	83
Varieties of pipettes	84
Price list of pipettes	85
The burette, a modification of the pipette	86
Price list of burettes	86
Toys founded on the principle of the pipette	87
The mysterious funnel	87
The magic can	88
Houdin's inexhaustible bottle	88
THE SYPHON	89
Great variety of forms	89
Action of the syphon explained	90
Toy syphons	92
Tantalus's cup	92
Hempel's syphon	92
Eye fountain	93
Price list of syphons	93
GLASS SYRINGES	93
GLASS MODELS OF PUMPS	94
Lift pump, a glass model	94
Force pump, a glass model	94
Force pump, another glass model	95
Fire engine, or double force pump, in glass	95
Fire engine, in metal	95
MODELS OF PUMPS, GLASS, BRASS MOUNTS	95
Valves, five varieties	95
Lift pump, glass and brass	96
Force pump, glass and brass	97
Pump stand, mahogany	97
Tate's school lift pump	97
Why water rises in pumps	97
SPRINGS AND FOUNTAINS	97
Classification of springs and fountains	97
Fountain produced by a fall of water under common atmospheric pressure	98
when condensed air forces a jet into free air	99
a. Glass fountain	100
b. Large zinc fountain	101
c. Small brass fountain	101
d. Set of four jets	101
e. The chromatic fire-cloud	102
Fountains produced when un-	

CONTENTS.

	PAGE		PAGE
FOUNTAINS —		Heron's fountain, in metal ..	105
condensed air forces a water jet into an exhausted receiver	102	The syphon fountain	105
		Intermitting spring, in glass	107
Fountains in vacuo	102	Intermitting spring, another variety	107
Fountain in vacuo with transfer plate	102	WATERWORKS	108
Fountain in vacuo, without transfer plate	102	Water-wheels, undershot, overshot, and breast-wheels ..	108
Heron's ball	103	Archimedian screw, in metal	109
Heron's fountain, in glass ..	104	Archimedian screw, in glass	109
Heron's fountain, another pattern	104	Appold's centrifugal pump ..	109

PNEUMATICS.

	PAGE		PAGE
AIR-PUMPS	111	Exhausting syringes	127
Varieties and powers of air-pumps	111	Condensing syringes	128
		Exhausting and condensing syringes	128
Pump with two vertical barrels	114	Ditto, with clamps	128
Another, of larger size, with raised plates	115	EXTRA FITTINGS FOR TATE'S PUMP	128
Tate's single-barrel pump on solid clamp	115	Extra joint with screw ..	128
		Arm to carry a syphon-gauge	128
Details of the construction of Tate's pump	116	Extra pump plates	129
Comparison of Tate's pump with pumps having double barrels	119	Connecting tubes, &c., for separate plates	129
		Syphon gauges, three kinds	129
		STOPCOCKS, CONNECTORS, AND FITTINGS FOR OCCASIONAL USE	130
Experiment of the freezing of water	120	
Experiment of the fountain in vacuo	120	Stopcocks	130
		Connectors, with 1, 2, 3, or 4 screws	131
Tate's pump, mounted on a pedestal	121	Caps for jars	132
Tate's pump, mounted on a table support	122	Washers	132
		Joints and connectors for tubes	132
Extra pump-plate	122	Blocks for a table	133
Tate's air-pump of large size	123	Ground-plate and hook ..	133
Air-pump with three barrels	124	Plate and sliding rod	133
Large Tate's air-pump, moved by circular action with fly-wheel	124	Clip and weight	133
		Single transfer plate	133
		Experiments with it	133
Small single-barrel air-pumps	126	Double transfer plate	134
Comparative labour that attends the working of these pumps	126	Aurora Borealis apparatus	135
		Clamps and keys	135
		Tallow-holder	135
Care of an air-pump	127	Grooved plate to secure draught of air	136
AIR-SYRINGES	127	

EXPERIMENTS ON THE PROPERTIES OF AIR.

In Four Groups.

	PAGE
Group A.—On the Weight and Resistance of Air	136
Estimation of the specific gravities of gases	136
Description of apparatus ..	137
Weight of air proven	138
Comparative weights of hydrogen and carbonic acid gases	139
Balloons	139
Baroscope, or cork and balance weight	140
Water hammer	141
V-formed water hammer ..	141
Fork-shaped water hammer ..	142
Guinea and feather experiment	142
Brass-work for it	142
Glass cylinders for it	142
Long tube for it	143
Windmill	143
Group B.—On the Expansion of Air	144
Bulb gauge and glass jar ..	144
Pair of narrow bottles for testing exhaustion	145
Lungs glass	146
Extrication of air from various bodies	146
From spring-water	146
From the pores of plants ..	146
From cork	147
From beer or ale	147
From an egg	147
From coke	147
Elasticity of air in an egg ..	147
Fruit and taper stand	148
The apple cutter	148
Bladder experiments	148
Preparation and care of bladders	148
Expansion of caoutchouc balls	149
Bladder frame and lead weight	149
Glass breaking squares ..	150
Cage for use with ditto	150

	PAGE
Paper smoke-jacks	150
Bells to be rung in exhausted receivers	150
Bell experiment that fails ..	150
Experiment that succeeds ..	151
Leslie's apparatus for freezing water	152
Evaporation in vacuo	153
Group C.—On the Pressure of Air	154
Downward pressure of the air	154
Filter cup for mercury shower	154
Shower of air in water	155
Hand glass	155
Bladder glass	156
Weight of the atmosphere ..	157
Crushing power of atmospheric pressure	157
Upward pressure of the air ..	157
Inverted glasses of water ..	157
Bottle with perforated bottom	158
Syringe and lead weight ..	158
Three-globes experiment ..	159
Fire syringe	159
Bottle imps, or Cartesian devils	160
Magdeburg hemispheres ..	161
Diving-bell	162
Group D.—On the measurement of Atmospheric Pressure ..	163
THE BAROMETER	163
Prices of fittings for barometric experiments	163
The Torricellian experiment	163
How to fill a barometer tube with mercury	164
Support for a barometer tube	164
Scale to measure the height of the mercury in the barometer	166
Syphon barometer	166
Deduction of the weight of the atmosphere from the Torricellian experiment ..	167

CONTENTS.

	PAGE
Mercury can be raised in a barometer tube by atmospheric pressure	167
Mercury is supported in the barometer by the pressure of the atmosphere, and it sinks in the tube when that pressure is removed	168
The elasticity or spring of the air is equal to its compressing force	169
Mariotte's apparatus; his law	169
Mariotte's apparatus constructed to show that under the pressure of two atmospheres air is compressed into half its ordinary bulk	170
Mariotte's apparatus constructed to show that under the pressure of half an atmosphere, air expands to twice its ordinary volume	171
GLASS RECEIVERS FOR AIR-PUMPS	172
Scale of dimensions and prices	174
Flat receivers for air-pumps	174
Cylindrical receivers	174
Bell-shaped receivers	175
Bell-shaped receivers with necks	175
Cylindrical receivers with necks	176
Conical fountain glasses ..	176
Guinea and feather glasses ..	177
Tall plain jars	177
Tall cylindrical receivers ..	177
Miscellaneous glass fittings..	178
INDEX	181

MECHANICS.

THE MECHANICAL POWERS.

1. IN the explanations of the properties of the Mechanical Powers, usually given in elementary works, it is taken for granted that rods, poles, planes, ropes, &c. are destitute of weight, free from roughness, from adhesion, and some other physical properties. But when the student comes to handle the apparatus, and to perform experiments with it, he finds that he must deal with all these impediments; and it is our business, in 'Scientific Handicraft,' to show him the means of overcoming them, and thereby rendering the experiments successful.

We begin by describing some pieces of apparatus that serve to facilitate operations with pulleys, levers, and other mechanical powers.

2. *Suspension Board*, or *Black Rail*, by which levers, pulleys, and other mechanical models are to be suspended from the top of the usual schoolroom blackboard. Size of the black rail: 4 feet long, 1 inch thick, 4 inches broad, with two brackets to hang it on the blackboard, and brass pins and holes to receive the other apparatus. In the following figures of levers and pulleys, the suspension rail is marked C D. Figure 2, price 5s.

The schoolroom blackboard should measure 4 feet in width to fit the rail, and 4 feet from top to bottom to give room for chalking upon it the results of the experiments made with levers, pulleys, cords, &c.

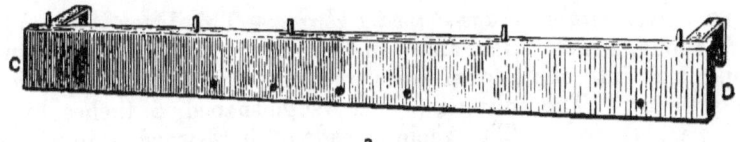

2.

3. *Pair of Iron Hooks* to keep the schoolroom blackboard in an upright position on the easel, in order to permit the apparatus to hang truly vertical from the suspension rail. 2s.

MECHANICS.

4. *Light Brass Scale Pans*, fig. 4, for holding weights and counterpoises, in experiments made with levers and pulleys. See Nos. 20 to 37. *Set of three pans in a box, 2s.*

5. *Weights.*—A set of 16 zinc weights, each $\frac{1}{10}$ lb. avoirdupois. Four of them have wires of suspension (fig. 6), the others have slots (fig. 7), so that they may be placed on any one of the wires to increase the weight. See figs. 6, 7, and 10. *Set of 16 weights, in two boxes, 7s. 6d.*

6, 7. These weights are to be used in all experiments with levers, pulleys, and other mechanical powers. Whenever, in the following descriptions and figures, the word *weights*, or the contraction *wts.* is used, it refers to these weights of $\frac{1}{10}$ lb. each.

8. *Iron Pins and Hooks* of two kinds (fig. 8 A and B), for attaching pulleys, &c. to the suspension board, &c. The wire of these pins is $\frac{1}{8}$ inch diameter, and they have sharp points. *Per dozen in a box, 1s.*

9. *Cord* fit for use with the pulleys, *per packet, 1s.*

THE LEVER.

10. *Lever with a Wedge-shaped Fulcrum.*—This lever (fig. 10) is 4 feet long, $\frac{5}{8}$ inch thick, and 2 inches wide. It is divided in the upper side into eight equal parts, and has eight brass suspension rings below. The fulcrum (f) is wedge-shaped, 6 inches high, and 2 inches wide. The whole is made of hard wood, stained oak colour. *Price, without the weights shown in the figure, 3s. 6d.*

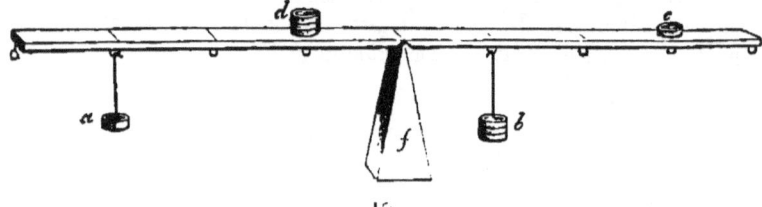

11. The weights of $\frac{1}{10}$ lb. each (No. 5) are used with this lever, and may be either placed on the upper part of the lever, as

shown by *c, d*, or suspended from the under side of the lever as shown at *a, b*; the weights with slots, No. 7, are used in one case, and those with wires, No. 6, in the other case. Equilibrium is produced by applying more or fewer weights, according to the respective distances of the weights from the fulcrum or point of support. Fig. 10 shows two sets of weights in equilibrium: the pile *d*, consisting of 3 weights of $\frac{1}{10}$ lb. each, placed on the *first* mark from the point of support, *f*, may be reckoned 30. The single weight of $\frac{1}{10}$ lb., *c*, placed on the *third* mark from the centre may be called 30; and these are in equilibrium. Then, referring to the suspended weights, the single weight, *a*, suspended from the *third* division of the lever may be called 30, while the three weights, *b*, suspended from the first division, may also be called 30; and these are therefore also in equilibrium. Any further weight, however small, added to these, oversets the equilibrium, and causes the lever to descend on the side that is overweighted.

12. If the lever, from becoming partially damp, or from some other cause, appears to be unequally balanced when placed upon the fulcrum, it can be justified by sliding a bronze halfpenny, or other flat piece of metal, along the surface till it reaches the required distance from the centre. The slots in the weights *c, d*, figs. 7 and 10, should be made to coincide with the lines that cross the lever.

If a small pair of scale pans, No. 4, are hung to the rings at the extreme ends of the lever, a balance is extemporised. If necessary, to give room for the pans, the lever can be raised higher from the table by putting the blocks, No. 70, or any other level piece of wood, below the wedge-shaped fulcrum.

14. Technically, one of the *forces*, that is to say, one of the weights or sets of weights applied to a lever, is called the *resistance*, or simply the *weight;* while the other force is called the *power*. The *weight* represents the work to be done: the *power* is the force that is to do the work.

15. *Lever for Suspension. Price 4s.*
This lever is made of hard wood stained. It is 3 feet long, and about 2 inches broad. It is divided on the front into 20 parts, and is provided with holes at all the divisions, and with four pairs of rings for supporting weights as represented by fig. 15. It is

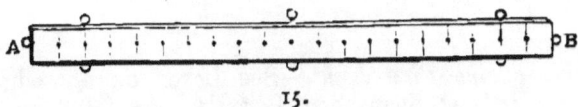

15.

to be suspended for use from the black rail in front of the school-room blackboard, as represented by figs. 17, 18, 19, 20, and with the help of the simple pulleys, No. 30, it serves as a lever of the first, second, or third kind.

After the lever is hung to the suspension rail, and before any weights are attached, it must be noticed whether it hangs horizontally or not. If defective, a bent wire must be put on it as a rider, and moved to the place where it justifies the level of the beam.

16. *Varieties of Levers.*—There are two kinds of levers, which differ essentially from one another; namely, those on which the forces act on *contrary* sides of the fulcrum, and those on which they act on *the same* side. They are, however, usually distinguished as of *three* kinds, according to the respective positions of the fulcrum, with reference to the situation of the power and of the resistance. In levers of the *first* kind the fulcrum is between the power and the resistance, as in figs. 10 and 17. In the *second* kind the resistance is between the fulcrum and the power, as in fig. 18. And, in the *third*, the power is between the fulcrum and the resistance, as in fig. 19; and the power exceeds the resistance as much as the distance of the resistance from the fulcrum exceeds the distance of the power from the same point. In other words, when a lever is in a state of equilibrium, the number of weights representing the power and the resistance, multiplied by their respective distances from the fulcrum, show identical results.

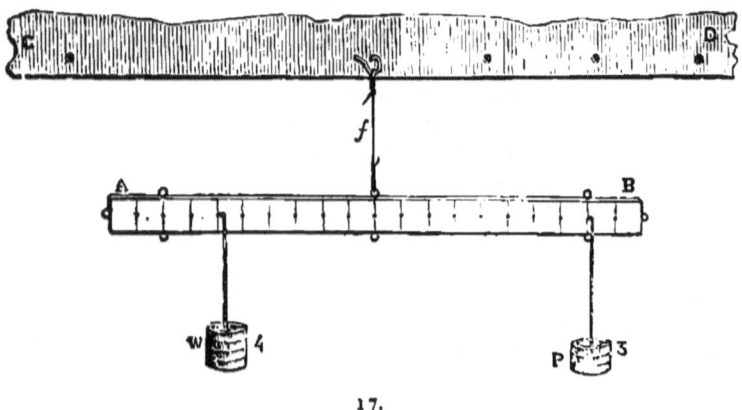

17.

17. *Lever of the First Kind.*—See Lever on a wedge-shaped fulcrum, fig. 10, and Suspended Lever, fig. 17. In this last example the lever is suspended by a pin and a cord, f, which constitutes the fulcrum. The power consists of 3 weights put on the 8th division of the lever, measuring from the centre, and the

resistance consists of 4 weights affixed to the 6th division of the lever, on opposite sides.

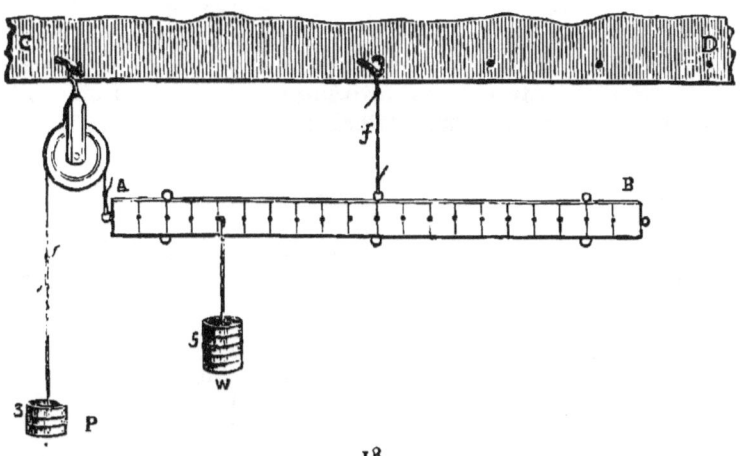

18.

18. *Lever of the Second Kind, fig.* 18.—In this case the lever is suspended by the cord *f* constituting the fulcrum; the power, consisting of 3 weights, is suspended by a simple pulley from the 10th division of the lever; and the resistance, consisting of 5 weights, is suspended from the 6th division of the lever, on the same side of the lever. As the 3 weights at division 10 pull the lever upwards, and the 5 weights at division 6 pull it downwards with equal force, the lever remains horizontal, although no force is applied to the arm B.

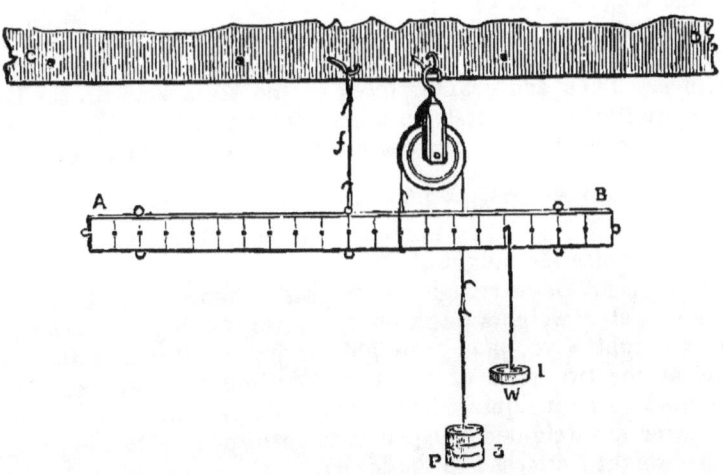

19.

19. *Lever of the Third Kind, fig.* 19.—As explained above, this kind of lever is not essentially different from a lever of the second kind. The power of 3 weights is suspended by a pulley from division 2, while the resistance of 1 weight acts upon division 6 of the lever. Hence the power is greater than the resistance, and is situated between the resistance and the fulcrum. They are brought into equilibrium by their respective distances from the fulcrum.

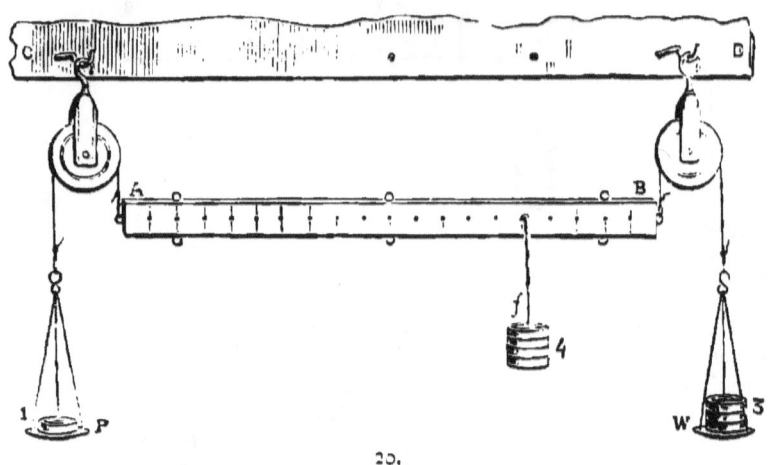

20.

20. *Gravesende's suspended Lever, fig.* 20.—This apparatus consists of—the suspension rail C D, the lever for suspension A B, a pair of simple pulleys (No. 30), 2 scale pans (No. 4), and a supply of zinc weights (No. 5).

21. Fasten the pulleys with pins to the suspension rail, and attach the scale pans to the lever after passing the strings over the pulleys. Then add counterpoises to the scale pans till the lever is in equilibrium and rests in a horizontal position. The lever is then in a condition to be sensible to the action of a very small weight.

Experiment A.—Suspend some weights at 0° in the middle of the lever. Half as many weights must be put into each scale pan to counterpoise the weight at 0°.

Experiment B.—Arrange the weights as represented by fig. 20; namely, put 4 weights = 40, on the lever, pulling it downwards, and 1 weight = 10, *plus* 3 weights = 30, in all 40, in the scale pans at the two ends of the lever, pulling it upwards. These weights will be in equilibrium. The scale pans at the two ends of the lever are weighted unequally, to correspond with the position of the weight *f* attached to the lever.

This apparatus shows in what manner a load can be placed upon a lever, so as to bear either equally, as in Experiment A, or very unequally, as in Experiment B, upon the men who support, or the horses that pull, the two ends of the lever.

THE PULLEY.

27. The *Pulley* consists of a wheel called a *sheave*, fixed in a frame or *block*, and turning on an *axis* or *axle*. Round its rim is a groove, on which a *cord* or band can pass to move it.

A *system* of pulleys is a number of pulleys acting in concert.

28. *Pulley Frame*, of black wood, with hooks, in which to stretch the four sets of pulley blocks, No. 30, 34, 35, 37, to prevent the entangling of their cords, when the pulleys are not in use, 3s. 6d.

29. *Sets of Pulleys mounted.*—The pulleys, or sheaves, are made of box wood, and are from $1\frac{1}{2}$ inch to $3\frac{1}{2}$ inches in diameter, the simple pulleys being $2\frac{1}{2}$ inches. The blocks are made of hard wood stained black, and are all provided with brass hooks. When they are to be used they must be suspended by means of pins and hooks, No. 8, to holes prepared in the suspension rail, No. 2, to receive them.

The effects of the friction of pulleys on their axles, of their rubbing against the blocks, the rigidity of the cords, and the weight of the pulleys, produce inaccuracies in experimenting, which have to be recognised and allowed for, in the manner described in the following articles.

30. *Simple Pulleys, a set of four*, made of box wood, the blocks of hard wood, stained black, all provided with brass hooks. Size of the pulleys $2\frac{1}{2}$ inches diameter. *Price of the set of four, 6s.*

31. *Experiments with Simple Pulleys.*—Fig. 31 represents one *simple pulley* attached to the suspension board so as to form a *fixed pulley*. The weights P and W, attached to the cord a, b, must be equal; the pulley is then at rest. If the power P is lowered a few inches, the weight W rises exactly as many. There is no *gain* of power by this pulley, but it affords useful aid in the *lifting* of heavy bodies, and in other modes of *transferring* power.

32. Fig. 32 represents a *Moveable Pulley*, B, combined with a *fixed pulley*, A, both attached to the suspension board. Before the weights P and W are added, it will be found that the apparatus is not in equilibrium, the moveable pulley B being heavier than the cord descending from A. In consequence, a scale pan, No. 4, must be connected with the cord a, and shot be added to counterpoise the pulley B. Then, upon adding weights, it will be found that 2 weights in the pan will balance 4 weights attached to the pulley

B, and that if the power P is pulled down, by extra weights or otherwise, say 12 inches, the weight w will rise 6 inches, the cords *b* and *c* each becoming shorter by 6 inches.

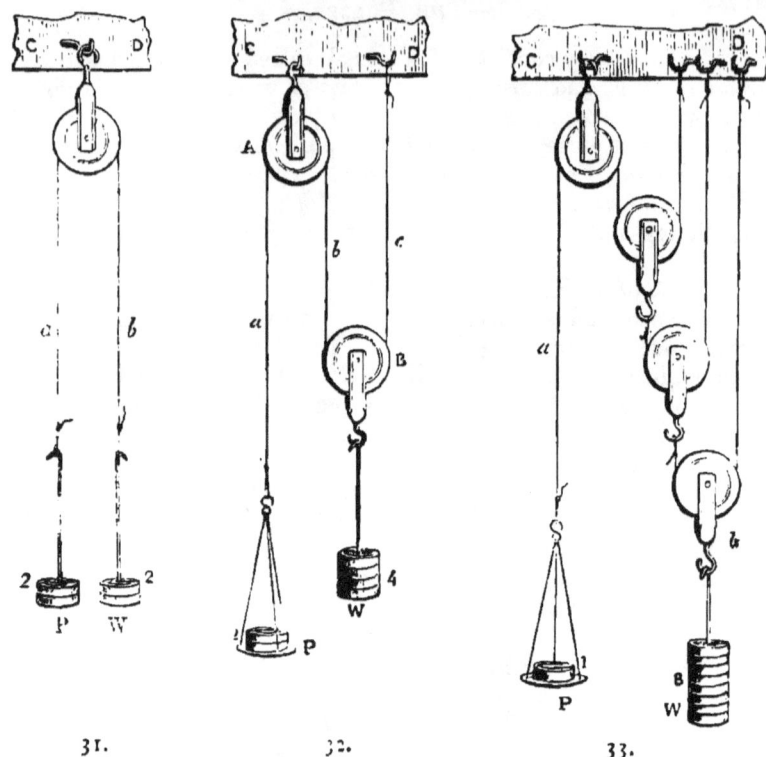

31.   32.   33.

33. *A System of Pulleys, fig.* 33.—In this system there are 3 moveable pulleys and 1 fixed pulley, and with such an arrangement we can, with a power of 1, counterbalance a resistance of 8. The fixed pulley merely affords a convenient mode of applying the power. After mounting the pulleys on the suspension board, as shown by the figure, the scale pan and any necessary counterpoise must be added to the cord *a*, before the weights are applied to constitute P and W. After the equilibrium is established, any extra weight, beyond the quantities specified above, added either to P or W will overset the equilibrium, and put the pulleys in motion. When P descends, it will pass over *eight* inches for *one* inch that W ascends.

34. *Pair of long Three-sheave Pulleys*, made of box wood, the blocks of hard wood stained black, provided with brass hooks. Size of the largest pulley, 2½ inches diameter. Fig. 34. *Price* 5s.

THE PULLEY. 9

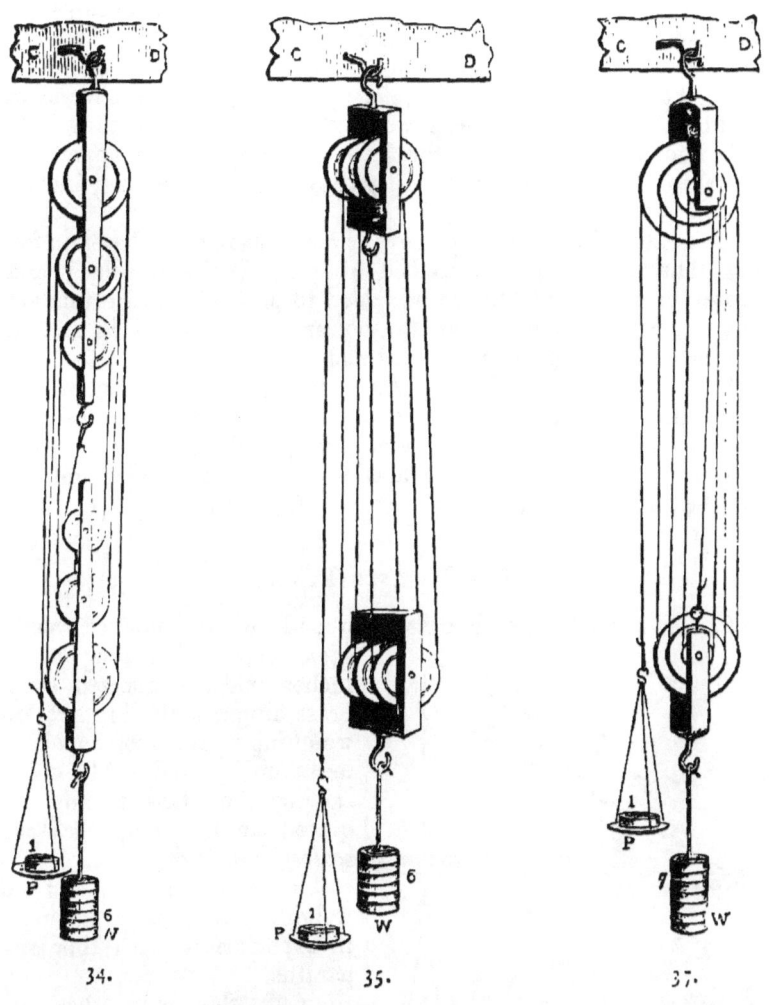

34.   35.   37.

35. *Pair of Square Three-sheave Pulleys*, made of box wood, the blocks of hard wood stained black, provided with brass hooks. Size of the pulleys, 2¼ inches diameter. Fig. 35. *Price* 5s.

36. The two systems of pulleys, Nos. 34 and 35, both have six cords to support the weight, and one cord for the power. In each case the weight supported is six times greater than the power, consequently for *every inch* that the weight w rises the power p descends *six inches*. In mounting these systems of pulleys for experiment the scale pan and counterpoises must be used in each case, because the friction and weight of the pulleys is considerable.

37. *White's Concentric Pulley, fig. 37.*—The pulleys are made of box wood, the blocks of hard wood stained black, provided with brass hooks. Diameter of the largest pulley of the upper set, 3¼ inches; diameter of the largest pulley of the lower set, 3 inches. *Price* 8s.

The action of this set of pulleys is accompanied by less friction than that of the sets described above. When arranged and counterpoised, the weights to be added to produce equilibrium are 7 for the resistance and 1 for the power, and when set in motion the spaces described by the weight and the power are as 7 to 1.

38. *Pulley Frame, and Set of Brass Pulleys.*—Consisting of two brass pillars, a long cast-iron base, and a mahogany connecting beam at top. Size 2½ feet wide and 3 feet high. The brass pulleys are the same, in size and form, as the wooden pulleys described above, No. 30, 34, 35, 37. *Price of the set*, 4l. 4s.

The Inclined Plane.

46. *Inclined Plane*, with base, both made of hard stained wood; each piece 2 feet long and 3 inches wide, connected by a brass hinge; with a brass roller weighing 1 lb. avoirdupois, a brass pulley, and a block for raising the plane to any required angle. *Price, without scale pan and weights*, 9s.

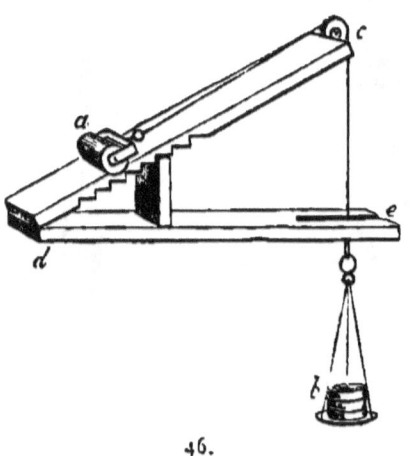

46.

The scale pan No. 4, and the weights No. 5, may be used in experiments with this apparatus.

47. In this, as in other mechanical powers, time is lost as power is gained; for the vertical height to which a body

is raised by means of the inclined plane is equal only to the height of the plane, while the space through which the power descends is equal to the length of the plane: the less the height of the plane, the greater the weight that can be raised on it by a given power. When a plane is twice as long as it is high, ½ lb. at *b*, fig. 46, acting over the pulley *c*, will balance 1 lb. at *a*, or anywhere on the plane, and so of all other quantities and proportions.

THE WEDGE.

48. *Wedge*: 6 inches long, 3 inches wide, with a divided cylindrical block, 7 inches long and 4 inches in diameter, bound with caoutchouc bands. Fig. 48. Price 4s.

49. The action of the wedge is that of a double moveable inclined plane, presenting two faces to two resistances to be overcome. The great practical advantage in the use of the Wedge, and that which gives to it its marvellous power, is, that it admits of being driven by *impact*, and that any force of *impact* is infinitely great as compared with any force of *pressure*. A second useful property of the wedge is, that it *retains* every new position into which it is driven between the resisting surfaces.

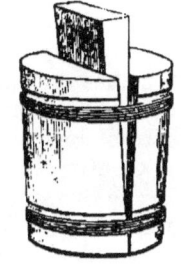

48.

THE WHEEL AND AXLE.

50. *Wheel and Axle*, consisting of two concentric pulleys cut out of one block of hard wood; the largest pulley 6 inches in diameter, the smaller 3 inches; with central pin for attaching the wheel and axle to the black rail, No. 2, and round which the pulleys move. Price 2s.

The weights No. 5 are to be used with this model.

51. This machine may be considered a kind of perpetual lever, having its fulcrum or prop in the centre of the axis and the wheel. The acting, or longer part of the lever, is the radius or half the diameter of the wheel, and the shorter, or resisting part, is the radius of the axis. In the model above described, the larger

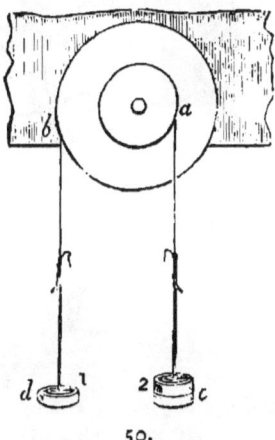

50.

radius is 3 inches and the smaller 1½ inch. Consequently, 2 weights suspended from *a* will be counterbalanced by 1 weight suspended from *b*. As to time, it follows of course, that when in motion, while the weight *c* ascends 1 foot the weight *d* must descend 2 feet.

This mechanical power is applied to many uses, forming the essential part of the winch, the windlass, the capstan, and the crane. Its power is increased by enlarging the larger wheel and diminishing the smaller; but this mode of gaining power is only practicable to a certain extent. When it is necessary to lift very heavy weights, the Compound Wheel and Axle, No. 52, is made use of.

52. *Compound Wheel and Axle.*—The wheel of this model is 5 inches in diameter; the thick axle 3 inches, and the thin axle 1½ inches diameter; the whole apparatus is 9 inches long, made of hard stained wood. Fig. 52.

53. *Bracket* by which the Compound Wheel and Axle is attached to the Suspension Rail, No. 2. This bracket serves also for use with the Model of the Screw, No. 56.

54. The Compound Wheel and Axle, No. 52; the Bracket, No. 53; the Model of the Screw, No. 56. *Price of the set*, 18s.

55. In the Compound Wheel and Axle the axle is made of different thicknesses, as at *a* and *b*, and a cord coils round these axles in different directions, and passes round a moveable pulley, *d*, from which is suspended the weight that is to be lifted, w.

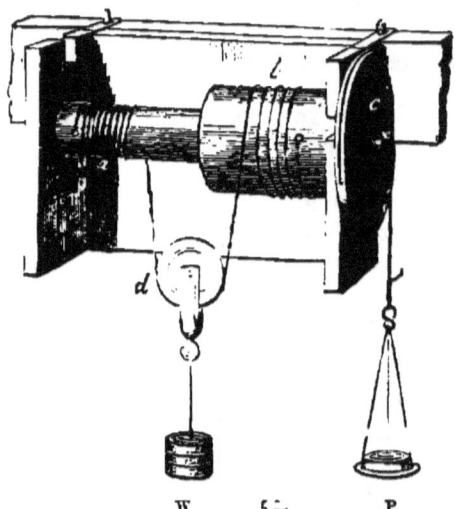

When the wheel *c* makes one turn, by the action of the power P, the cord is coiled up once round the large axle *b*, and is uncoiled once from the small axle *a*. Consequently it is shortened by a space equal to the difference of the circumferences of the axles *a* and *b*, and the weight w rises a space equal to half that difference. If the weight rises half an inch only while the cord round *c* descends 50 inches, then 1 lb. at P will balance 100 lb. at w.

The power and the rapidity of action of this machine depend

upon the difference of the diameter of the two axles. The less the difference of these diameters, the less the power required to maintain the weight in equilibrium, and to move it. By making the two axles more nearly of the same diameter we can diminish the power necessary to raise any given weight, or increase the weight which any given power will raise, without limit.

It is evident, that to cause the pulley d to descend, the action of the wheel c (or the winch handle used instead of it) must be reversed, upon which the cord will be unrolled from the wide axle and rolled upon the narrow axle.

The Screw.

56. *The Screw.*—Apparatus for illustrating the principle of the action of the Screw. Size of the screw 6 inches long, 3 inches diameter, with square bottomed thread made of hard wood stained. Mounted on a spindle with handle, and adapted to the Bracket, No. 53. *Price:* see No. 54.

57. The screw is an inclined plane coiled round a cylinder, as represented by fig. 56. There are two kinds of screw, the *male* screw, with projecting threads as here represented, and the *female* screw, which has hollow spaces corresponding with the threads of the male screw. The female screw is often called a *nut*, which is a block perforated by a screw, that fits any part of the male screw. Every turn of a screw carries it forward in a fixed nut, or draws a moveable nut along with it by exactly the distance between two turns of its thread: this distance, therefore, is the space passed through by the *Resistance*, while the *Force* moves in the circumference of the circle described by the handle of the screw. The disparity between these lengths or spaces is often as a hundred or more to one : hence the prodigious effects which a screw enables a small force to produce.

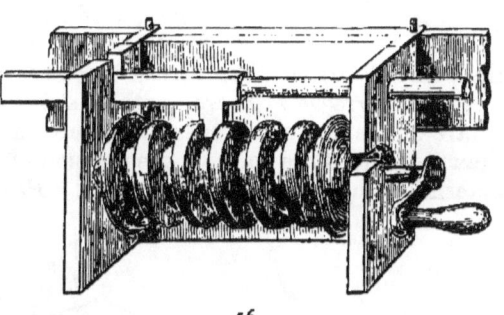

56.

---

58. *Collection of Working Models to illustrate the Mechanical Powers*, consisting of the Lever, the Pulley (several sorts), the Wheel and Axle, the Inclined Plane, the Wedge, and the Screw.

German manufacture, neatly made in white wood, with a printed and illustrated description. In a smooth box with sliding cover, size 12 by 18 inches. *Price* 21s.

This box of small models does not form part of the series above described.

Centre of Gravity.

70. Apparatus to illustrate the *Centre of Gravity*, namely, two equal *parallelopipeds* of hard wood, each 5¼ inches long, 4 inches wide, 2 inches thick, of a rhomboidal form. They stand firmly on end when separate, but, when placed on one another, they are on the point of falling: a penny piece suffices to overset the equilibrium. *Price* 2s.

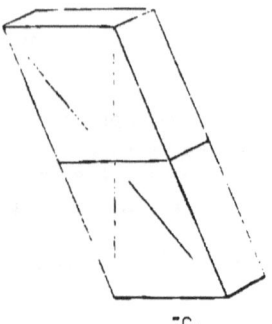

70.

They are marked with lines to show the centre of gravity.

71. *Leaning Tower*, or *oblique Cylinder*, 8 inches high, 4 inches diameter, in two pieces, made of hard wood. They stand firmly on end when separate, but are on the point of falling when placed on one another. *Price* 2s.

72. *An Irregular Board*, with a string, for showing how the centre of gravity of an irregular figure may be found. Lines are drawn on the board to show the position of the centre of gravity as determined by experiment. Fig. 72. *Price* 1s.

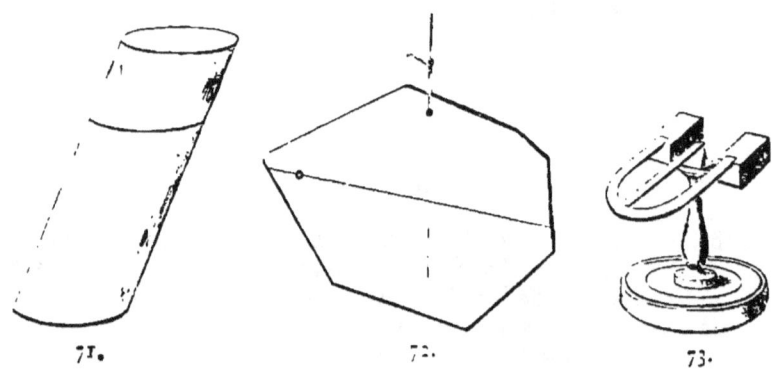

71.   72.   73.

73. *Apparatus* to show that the centre of gravity of a body is not necessarily within its volume, consisting of a *semicircle of brass*, weighted at the two ends, and supported on a brass stand, on a knife edge. Fig. 73. *Price* 3s. 6d.

74. *The Equilibrist*, fig. 74, small size, about 2½ inches long. Price 1s. 6d.

75. Ditto, large size, figure about 4 inches long. Price 2s. 6d.

76. The equilibrist is a child's toy, resembling fig. 74. It consists of a figure carved in wood, and reposing by a point on a little pedestal. Two wires fixed in the figure end in balls of metal. By this means the centre of gravity is thrown below the point of suspension, and the equilibrium is thus rendered stable. The figure is readily moved about, and can be made to oscillate freely; but after every movement it ends by placing itself in such a position of equilibrium that the vertical line passing by the centre of gravity passes also through the point of support.

74.

77. *The Bowl-About.*—This toy is founded on the principle of placing the centre of gravity very near the lower part of the figure, by accumulating the material in that region so as to produce stable equilibrium. Fig. 77. Price 2s. 6d.

78. " There is an amusing Chinese toy, which has the appearance of a little fat laughing man, sitting on the ground with his feet concealed under him; but where the feet should be there is only a rounded smooth surface, with heavy lead ballast placed in it, so low, as always, when allowed, to raise the body to the erect or sitting attitude. A child pushes the little fellow down again and again,

77.

and would persuade him to be still, but is surprised to see him always up the moment after, shaking about and as lively as ever."—*Dr. Neil Arnott.*

79. *Double Cone*, which apparently runs up an inclined plane. Fig. 79. Price 8s. 6d.

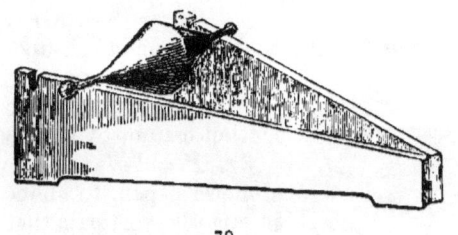

79.

This apparatus consists of a double cone and a sort of railway, the two lines of which depart from a centre and at the same time diverge and rise, as represented in fig. 79. When the double cone is placed horizontally across the inclined plane produced by these rails, it appears to run up the incline; but, in reality, the centre of gravity of the double cone descends between the widened rails, so that the cone sinks while it seems to rise. At the upper ends of the rails are notches, to hinder the cone from running off the rails.

DEGREES OF STABILITY.

80. *Cone for showing states of Equilibrium.*—The model measures 4 inches across the base, and 5 inches in height. Price 1s. 9d.

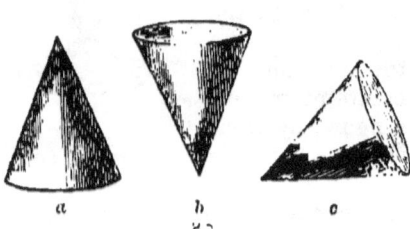

Fig. *a* shows stable equilibrium; fig. *b* shows unstable equilibrium; and fig. *c* neutral or indifferent equilibrium, the model being placed upon a horizontal plane.

81. *Inertia Apparatus*, by the use of which a card can be struck from under a ball that rests directly upon it, without propelling the ball. This experiment is intended to prove that a body which is at rest cannot put itself into motion. Price 5s.

For other illustrations of inertia, consult the article on the centrifugal machine, No. 97.

82. *Tin Pan and dried Peas.*— The pan, with peas loose in it, is lifted quickly upwards and the motion is then suddenly arrested, upon which the peas fly out of the pan. Price 1s.

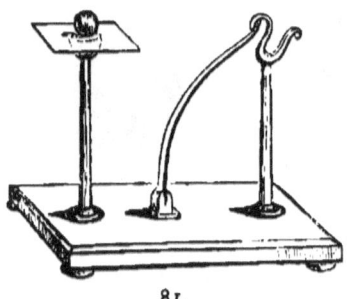

This serves to prove that a body which is in motion cannot of itself arrest its motion.

83. *Plumb-line*, used with mercury, held in a flat stoneware pan, to show by reflection that the line is strictly vertical, that is to say, perpendicular to the horizontal surface of the liquid metal. The line, 1s.; pan, 6d.

84. *The Pendulum.*—Apparatus for verifying the laws of the pendulum, consisting of four balls and cords, suspending them to

PENDULUM.   17

a frame, with arrangement for regulating the length of the cords. *Price* 12s. 6d.

The duration of the oscillations of a pendulum increases proportionally to the square root of its length, and this duration is independent of the substance, light or heavy, of which the pendulum is formed. The apparatus is contrived for the demonstration of these facts.

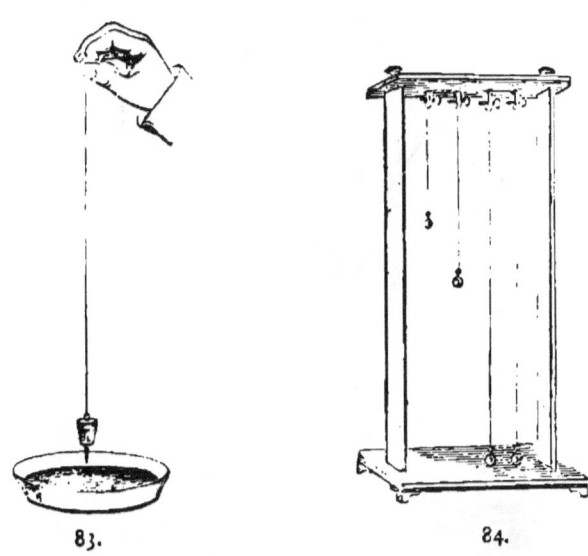

83.   84.

85. Fig. 84 represents a frame of black wood, from which are suspended four pendulums, a, b, c, d, the cords of which can be lengthened or shortened by means of the four screws to which they are attached. The shortest pendulum, a, has an arbitrary length, and accomplishes its oscillation in a given time. The first experiment to be made consists in giving to the cord of b such a length that its oscillation shall last twice as long as the oscillation of a, and to the cord of c such a length that its oscillation shall be three times as long as that of a. These results having been attained by trials, the cords are to be measured, when it will be found that the length of the pendulum b is four times, and that of the pendulum c is nine times that of the pendulum a. Consequently, the duration of the oscillations is in the direct ratio of the square roots of the lengths of the pendulums.

86. The second experiment is made with the fourth pendulum d. The length of this pendulum is made equal to that of pendulum c; but, whereas the balls of a, b, c, are made of brass, the ball of d is made of wood, of the same size. The oscillation of d is then to be

c

tried against that of *c*, when it will be found that the change of material in no respect changes the duration of the oscillations.

87. *The Parallelogram of Forces* can be demonstrated with the help of the black rail No. 2, the simple pulleys No. 30, the weights No. 5, and the scale-pan No. 4, belonging to this set of apparatus.

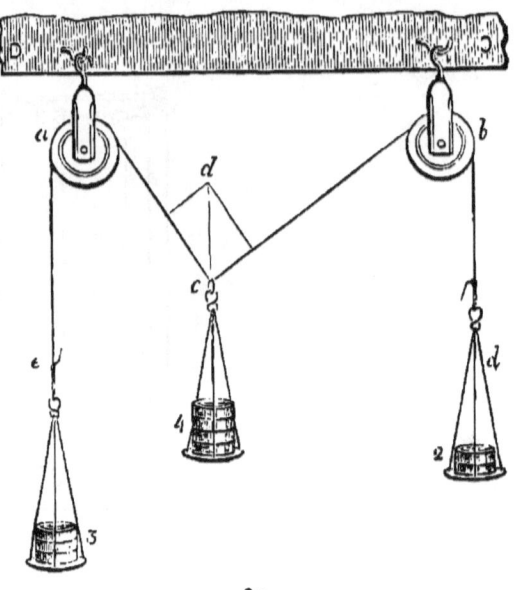

87.

Two simple pulleys, *a* and *b*, are to be fixed with pins on the suspension-rail c d, and a cord with a ring, *c*, in the middle, is passed over the pulleys. Two scale-pans, No. 4, are to be attached to the ends of the cord and one to the ring *c*, and these pans are to be brought into equilibrium by counterpoises; weights are then to be put into the two outer scale-pans, and other weights into the central pan. These weights must differ according to the angle to be produced. If the weights put into the outer pans are 3 and 2, as represented in the figure, then 4, put into the central pan, will effect equilibrium, and the cords will produce at *c* an angle *a c b*. If this angle is chalked upon the blackboard behind the lines, and parallel lines are drawn to complete the parallelogram *c d*, then the vertical diagonal from *d* to *c* will measure 4, and agree with the weight suspended below it.

When desirable, one of the pulleys can be adjusted lower than the other by using a wire or string to suspend it from the rail.

COLLISION. 19

88. *Adhesion Plates.*—A pair of well-polished circular glass plates, 4 inches diameter, ¼ inch thick. *The pair*, 4s.

90. *Toy, illustrative of Adhesion.*—The magic waltzers, a pair of coloured wooden figures mounted on a glass lens convex below. When placed on a glass plate, with an intervening drop of water, and the glass plate is slightly inclined, the figures revolve or waltz. Price 2s.

91. *Single Dancing Figure*, for the like purpose. Price 1s.

92. *Percussion Machine, for Experiments on the Collision of Elastic Bodies.*—A frame, with five ivory balls. Price 45s.

93. *Percussion Machine*, cheaper description, with ten stone balls, in a rectangular wooden frame. Fig. 93. Price 10s. 6d.

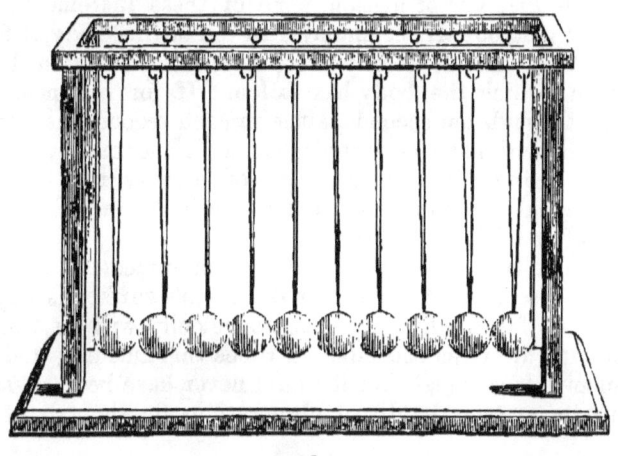

93.

"For making experiments on the phenomena of collision, it is most convenient to suspend the bodies employed by threads, in the manner of pendulums. When a ball at rest is struck by another equal ball, it receives a velocity equal to that of the ball which strikes it, and this ball remains at rest. And if two equal balls meet or overtake each other with any unequal velocities, their motions will be exchanged, each rising to a height equal to that from which the other descended. The effect of collision takes place so rapidly, that if several equal balls be disposed in a right line, in apparent contact with each other, and another ball strike the first of them, they will all receive in succession the whole velocity of the moving ball before they begin to act on the succeeding ones; they will then transmit the whole velocity to the succeeding balls, and remain entirely at rest, so that the last ball only will fly off.

"In the same manner, if two or more equal balls, in apparent contact, be in motion, and strike against any number of others placed in a line, the first of the moving balls will first drive off the most remote, and then the second the last but one, of the row of balls which were at rest; so that the same number of balls will fly off together on one side, as descended to strike the row of balls on the other side—the others remaining at rest."—*Dr. Thomas Young.*

Laws of Falling Bodies.

95. "*The Velocity which is communicated to a body falling freely by Gravity.*—Bodies falling freely, near the earth's surface, have communicated to them equal additions of velocity in equal times; and since, by the first law of motion, none of these increments of the velocity are lost, but all accumulated in the falling body, it follows, that its actual amount at any time, must be proportioned to the time during which the body has fallen. If, for instance, a body has fallen through ten seconds, since in each second the attraction of the earth will have communicated to it the same addition of velocity, and since all these additions of velocity will be retained in it, its actual velocity must be ten times that which it would have had after falling one second.

"The velocity which gravity thus communicates to a falling body in *each* second of time near the earth's surface is $32\frac{1}{6}$ feet; so that, after falling five seconds, its velocity will be five times this amount; after ten seconds, ten times this amount; and so on. This velocity is so great that it would never have been possible to ascertain its amount by direct observations on the fall of heavy bodies.

"Could we, however, by any contrivance *neutralise* the gravitating tendency of a body to any known amount—reduce it, for instance, to *one-half*, or *one-tenth*, or *one-hundredth* of what it was—since we should diminish the *velocity*, communicated to it in each second, precisely to the same amount, we might thus render its motions so slow, that they might be *observed* and *measured;* we might thus find the amount of the additional velocity actually communicated to it in each second, and this multiplied by the known number of times by which we had previously diminished the force of its gravity, would give us the velocity which that fall would communicate in each second, when *undiminished*. This is the object of *Attwood's* machine."—*Moseley,* 'Illustrations of Mechanics,' p. 309.

96. *Attwood's Fall Machine,* with large graduated pendulum and large brass pulley, with axle mounted on friction wheels, double perforated stage, and stop stage, completely fitted, with centimetre graduations, and extra graduations for showing velocities equal to

the square of the times. Two buckets of brass, of equal dimensions and weight, connected by a fine silk thread, and three brass weights, two of the form shown at the bottom of fig. 96 and one of square form—one of the long weights, however, being equal in weight to both the others combined —there is an arrangement at the back of the upright support in connection with the pendulum which causes the weighted bucket to commence falling the moment the pendulum is set in motion. The base board of the instrument is supplied with levelling screws. When in use it is requisite to observe that all the wheels move freely and are free from dust or hardened oil; also the bearing of the pendulum must be examined for the same reason. The rod leading from the pendulum, with its cross piece fitted with an india-rubber stop, must be so adjusted with the thumb-screw attached for that purpose, that when the pendulum is at rest the rubber may just touch and arrest the motion of the pulley wheel, but at the instant the pendulum is set in motion the stop must be lifted off the wheel to allow it free action.

The stages can be placed at any part of the scale most convenient for the experiment required; the length of the pendulum is to be adjusted according to the weight and velocity of the falling body or else to strike seconds on the bell, and then firmly fixed by the thumb-screw in the necessary position.

The heaviest of the 3 weights serves to illustrate the Law of Falling Bodies that the velocity is equal to the square of the time. If, for instance, the weight be adjusted to 0, and the pendulum set in motion at the first stroke, it will fall to 1, at the second to 4 (No. 8 on the scale), at the third to 9 (No. 13 on the scale), &c. &c., till the ninth to 81 (No. 162 on the scale); the first series of these numbers is marked on the outer graduations.

This experiment, as well as the following and many others which can be made with this machine, will be found described in every treatise on Physics.

The second experiment for illustrating uniform velocity by using the two lesser weights, the smaller of which is termed the inertia weight, for counteracting the inertia of the mass of matter to be moved, namely, the pulley wheel, friction wheels, pair of buckets and weights.

The perforated stage is placed at one of the numbers on the outside scale showing squares of times. On starting, the weight at first falls according to the first experiment in velocities equal to the squares of the times, but at the instant the bucket passes through the perforated stage the longer weight will be removed, and the bucket containing then only the inertia weight will continue to descend at the uniform rate of the velocity acquired at the time of the removal of the longer weight, and not with accelerated velocity as in the first experiment.

Fig. 96 represents the general form of this machine, but is not accurate in details; in particular, it shows the graduation very imperfectly, and omits entirely the representation of the friction wheels which support the pulley at the top. *Price of the machine* as described above, 10*l*. 10*s*.

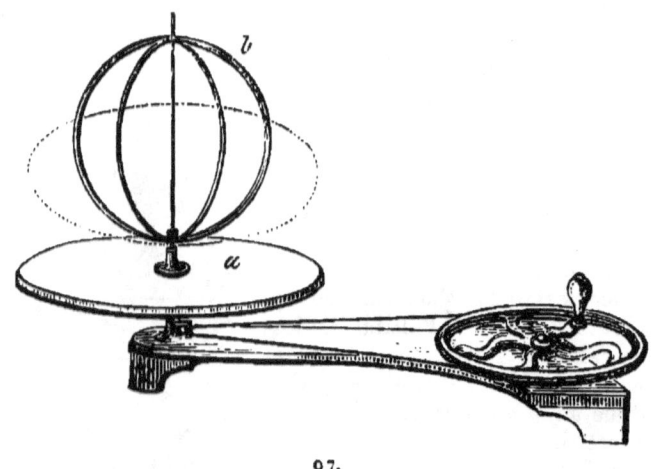

97. *Whirling Table, or Centrifugal Machine*, with Adjuncts for 12 experiments to illustrate the Laws of Central Force. *Price of the machine, with* 12 *Adjuncts*, £1. 6*s*.

CENTRIFUGAL MACHINE.

97 A. *Price without the* 12 *Adjuncts, but with the Table* a, 3*l.*
The Adjuncts separately, at the price affixed to each.

The machine is represented by fig. 97, and the adjuncts by figs. 97 A to 97 M, of which pieces, Nos. 97 A and 97 B are shown as if screwed on the machine and in action.

The Centrifugal Machine consists of an iron frame or base, on which are mounted an iron wheel to produce rotation, and a pulley for supporting the adjuncts, and conveying the motion to them for the various experiments. The wheel and pulley are connected by a stout band of catgut, which can be strained when necessary by a screw placed under the wheel. The adjuncts represented by figs. 97 A to 97 M are all mounted on female screws, which fit the male screw affixed to the upper end of the axle of the pulley. All these pieces of apparatus, A to M, can be readily screwed on the machine for experiments, and be rapidly removed when the experiments are finished.

Fig. 97 A.
A circular table of black wood, 12 inches diameter, which being put on the pulley (see *a*, fig. 97) acts as a fly-wheel to steady the action of the other pieces of apparatus.

Fig. 97 B.   *Price* 6s.
This represents an apparatus formed of two elastic rings of thin brass, which are fastened at the bottom to a vertical iron rod, and at the top to a ring which slides freely upon the rod. The lower end of the rod is screwed to the pulley. When the centrifugal machine is set in motion the rings represent a sphere, but when the whirling of the table becomes rapid the rings assume the form shown by the dotted oval. This experiment is made to illustrate the tendency of spherical bodies, which revolve rapidly on their axes, to assume a spheroidal form—as the earth does.

Fig. 97 C.   *Price* 5s. 6d.
An apparatus to exhibit the effects of inertia. It is to be screwed above the blackboard, fig. 97 A, and rotated slowly.

Observe the effects: the ball does not immediately begin to move with the board, but endeavours to continue in its state of rest. As the rotation proceeds, the board communicates its own motion to the ball, and shortly the ball keeps a position on the board, assuming the same velocity

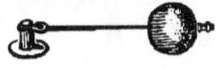

97 C.

as the board, and having no relative motion of its own. The apparatus now exhibits the state of matter common on the surface of the earth, where moveable things of all kinds retain their places, notwithstanding the rapidity of the earth's rotation.

But to pursue the experiment in hand. Stop the rotation of the board with your hand. The ball does not stop when the board is stopped, but continues its revolutions on the board, until it is stopped by friction and the action of the air. This shows that matter, once put into motion, continues to move until it meets with resistance.

Fig. 97 D. *Price 7s. 6d.*

This figure represents a couple of brass balls, connected by a

97 D.

fine wire, or a thin metal tube, and which slide upon a long wire affixed to a frame, upon the bottom of which is a screw adapted to the screw of the whirling table. When these balls are placed in such a position on the long wire that their common centre of gravity is directly over the axis of rotation, then, upon whirling the machine, the centrifugal force of the two bodies will be equal, and they will continue to rotate round each other. But if the common centre of gravity of the two balls be removed to either side of the axis of rotation, the system will fly off in the same direction when the apparatus is rotated.

Fig. 97 E. *Price 6s. 6d.*

This figure represents a metallic ring, from the upper part of

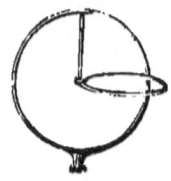

97 E.

which a smaller metal ring is suspended by a metal bar. At the bottom of the large ring is a screw adapted to the pulley of the whirling table.

When this apparatus is at rest, the small ring hangs down from the end of the bar. But when the apparatus is whirled by the machine, the little ring rises into a horizontal position and rotates rapidly.

Fig. 97 F. *Price 8s.*

Fig. 97 F represents an apparatus with three pendulums, to

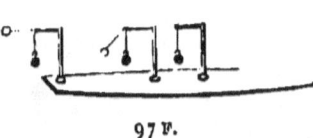

97 F.

show the effects of the centrifugal force according to the position of each pendulum in reference to the centre of motion. The screw that connects this adjunct with the whirling table is situated in the centre of the foot board or bearer, directly under the first pendulum on the right hand. The blackboard, 97 A, must be used with this piece of apparatus.

When the machine is put into gentle motion, the pendulum which hangs over the axis of rotation, retains its vertical position unchanged; the second pendulum flies off at an angle of 45°; the third pendulum flies out horizontally. These positions are shown by the dotted lines in Fig. 97 F.

Fig. 97 G. *Price* 5s.

This apparatus consists of a glass globe about 3 inches in diameter, mounted on a brass foot, by which it can be screwed to the axle of the pulley of the whirling machine. It should be a little more than half filled by water coloured red.

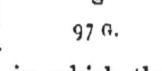

When the apparatus is rotated, the water, obeying the centrifugal force, first rises up on all sides of the globe, and, when the rotation is rapid, it quite leaves the bottom of the glass and forms coloured bands round the sides of the globe. It has been assumed that this experiment illustrates the manner in which the belts of the planet Jupiter are produced.

The globe, fig. 97 G, may be filled with water, except a small space, and its mouth be closely corked. When rotated in that state, the air forms a spherical bulb in the midst of the mass of water.

Fig. 97 H. *Price* 10s. 6d.

This apparatus is formed of two glass tubes mounted in brass and on a wooden bearer. Both tubes can be unscrewed and opened for the insertion of substances for experiment. One of the tubes is to be half filled with water and then have nearly half as much mercury by measure added. The other tube is to be half filled with water,

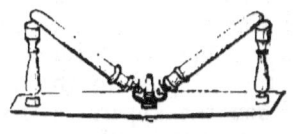

and then a small cylinder of cork is to be added. This cork must be narrow enough to slide easily in the glass tube, and about 1½ inch long. When the apparatus, thus prepared and screwed together, is at rest, the mercury falls below the water in one tube and the cork rises above the water in the other tube. But when the whirling machine is put into action the centrifugal force overcomes the force of gravitation; the mercury rises through the water and flies to the extreme end of the tube on one side, while, on the other side, the cork sinks through the water which rises to the extreme end of the tube. *The tubes should be emptied and cleaned after each experiment.*

Fig. 97 I. *Price* 15s.

This figure represents a model of James Watt's Governor, or centrifugal regulator for the steam engine. This apparatus is fully described in all works on Heat, or on the steam engine.

Fig. 97 K. *Price* 3s.

This figure represents a brass groove, or gutter. Three marbles are placed in it, the middle one of different colour from the other two. The apparatus is screwed upon the whirling table, and rotated slowly. The central marble remains at the bottom of the groove, being just on the axis of rotation. The other two marbles fly up to the centre of the bend, one on each side, as shown by the figure. If the whirl increases in rapidity the groove swells out, and the marbles remain there. Sometimes, however, an irregularity in motion causes the marbles to fly out of the groove. To prevent this, it is prudent to put corks in the extreme ends of the bent groove.

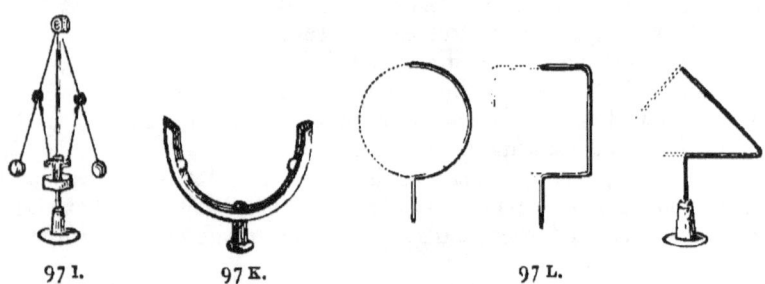

97 l.   97 k.   97 l.

Fig. 97 L.   *Price 4s.*

Fig. 97 L represents three bent and polished brass wires. There is one foot, which serves for all the three wires, and it contains a screw adapted to the axis of the pulley of the whirling table. When the apparatus is whirled with rapidity the wires represent the figures marked in outline; that is to say, one represents a cone, another a cylinder, and the third a globe. If the motion is very rapid the wires swell out at the centre, and the sphere becomes a spheroid.

Fig. 97 M.   *Price 8s.*

This figure represents an apparatus for investigating some of the laws of centrifugal force. The reader will find explanations in some work on the theories of Mechanics and Physics.

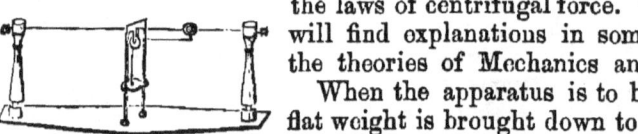

97 M.

When the apparatus is to be used, the flat weight is brought down to the bottom of the upright wires with which it is connected. That brings the ball back to the centre of the horizontal wire. The apparatus is screwed to the whirling table and rotated, upon which the ball flies to the end of the long wire and pulls up the flat weight.

When the experiments with this apparatus are to be made the subject of investigations extra weights are required, and the whole of the weights are to be made with relations to one another.

98. *Atmospheric Optic Marvel.* — A collection of toys illustrating the effects of centrifugal motion. Fig. 98. *Price, in a box,* 2s. 6d.

This toy consists of a metal cylinder, into which air is blown by the mouth through a flexible tube. The air escapes by small holes which act on a series of fans, and drive a circular table with great velocity in a horizontal position. There are some bent and coloured strips of metal which can be mounted on the table and driven round with such swiftness as to give the appearance of solid bodies. Thus the slip shown on the figure represents a perfect vase; others represent a hat, a water jug and basin, a tea-cup and saucer, a lamp globe, &c. When the air is blown pretty strongly, the figures swell out at the sides as represented in fig. 98, and as described in reference to the centrifugal globe, fig. 97 B.

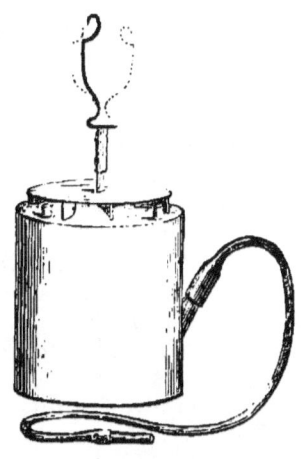

98.

101. *The Gyroscope, or Rotating Apparatus*, to show that when a body rotates its weight apparently diminishes, with several additions, to show the experiments of Bohnenberger, Fessel, Plücker, and others on inertia, &c. Price 35*s*.

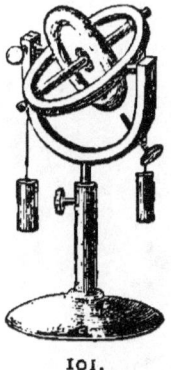
101.

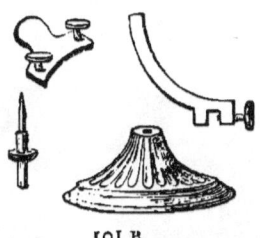
101 B.

This instrument is designed to show a variety of experiments illustrating inertia and rotary motion.

Inertia, which all matter possesses, is rather a passive property than an active force, and is manifested by matter when in motion as well as when in a state of rest; in the one case signifying that matter is incapable of moving of itself when in a state of rest, or of stopping itself when in a state of motion, and

28   MECHANICS.

also in resisting any attempt to change the plane in which matter is moving.

The Gyroscope in its complete form, made in gun metal and finely finished, is shown in fig. 101; with weights, pivots, extra screws, &c. fig. 101 B. Each instrument is accompanied by a printed description of its working details, and particulars of the experiments which can be performed with it.

102. *Gyroscope*, fig. 102. Gun metal, on wooden stand, *price* 7s.

105. The *Gyroscopic Top*, or *Top of Tops*.—A set of 3 coloured iron tops that spin either separately, or upon one another. *The set in a box*, 2s. 6d.

Directions.—The string should be wound tightly round the straight part on the under side, the first round being tightly drawn into the groove to prevent it from slipping off.

The large top should be spun first and only gently; the second size top should then be spun as forcibly as possible, and dropped carefully into the cup of the large top. The smallest top should be spun in the same way. The large top will soon take up the momentum of the smaller ones, and may be kept up for hours by re-spinning the smaller ones and placing them upon it. The tops spin best when the points are slightly oiled. Further details are given in a printed note sold with the tops.

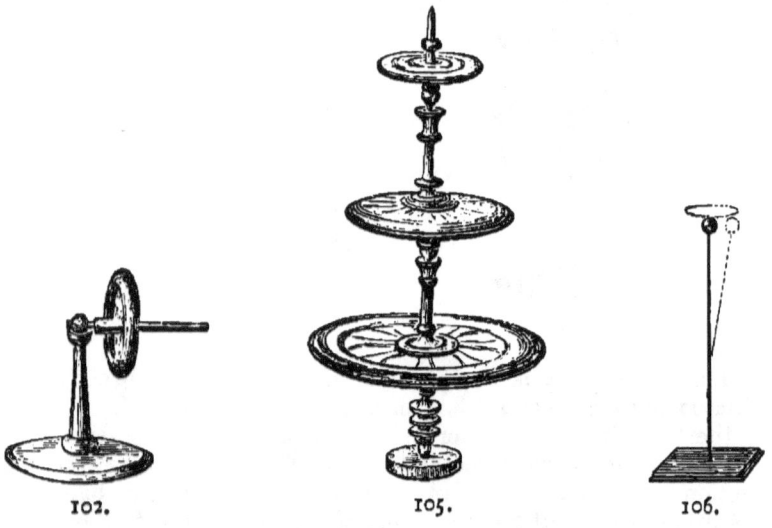

102.   105.   106.

106. *Vibrating Wire.*—Fig. 105. *Price* 1s. 6d.

A steel rod, 13 inches long, mounted on an iron foot, with a polished steel bead at the top.

CENTRIFUGAL RAILWAY.

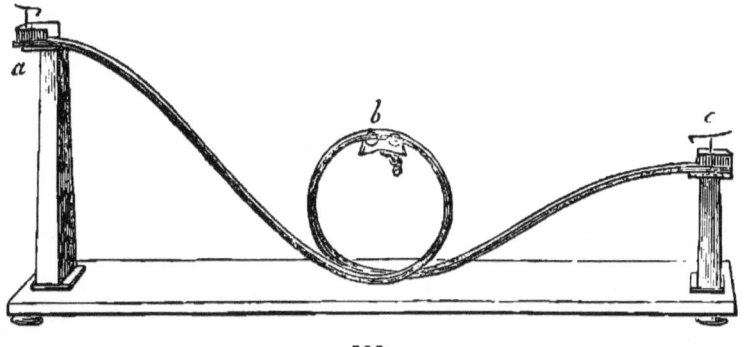

107.

107. *Model of a Centrifugal Railway.*—Fig. 107. *Price* 18s.
This apparatus serves to show the effects of centrifugal force. The form is that of a slightly-made iron railway, which rises to a spiral in the middle, and the two ends of which rise to different heights. A waggon is made to depart from the higher level at *a*, and, apparently in spite of the laws of gravity, it passes round the spiral, in the middle being head downwards at *b*, and thence goes up the second incline and rests at *c*, the station at the summit of the second column.

108. *Constructive Mechanics.*—Models for illustration, each in two pieces of stained wood. *Price of the set of four pairs*, 6s.

1, Half-lapping; 2, Dove-tailing; 3, Scarfing; 4, Tenon and Mortice.

109. *Parallel Motion.*—Two wooden rods, each 12 inches long, with a 5-inch connecting link, and two brass pins at *c* and *d* for centres. These pins are to be fixed on the suspension-board. A hole in the centre of the link enables the lecturer to insert his chalk-pencil and demonstrate that the motion is a straight line. Price 2s. 6d.

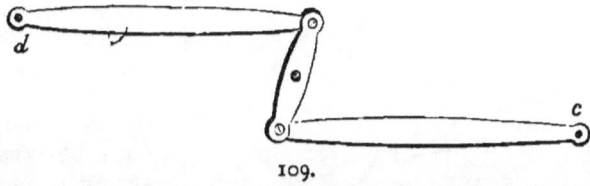

109.

110. *Intermittent Motion*, consisting of an escapement-wheel, with an escapement made of stained hard wood, mounted on a black-board, measuring 15 inches by 8 inches, with brackets for suspending it to the black rail. Price 7s. 6d.

111. *A Train of Wheels.*—The large-toothed wheel, A, 7½ inches

in diameter, with 60 teeth, and with 1½-inch pulley; the small-toothed wheel, B, 2½ inches diameter, with 20 teeth, and with 5-inch pulley. The pair of wheels made of hard stained wood, mounted on a blackboard, measuring 18 inches by 10 inches, with brackets to suspend it to the black rail, No. 2. *Price, without weights,* 12s.

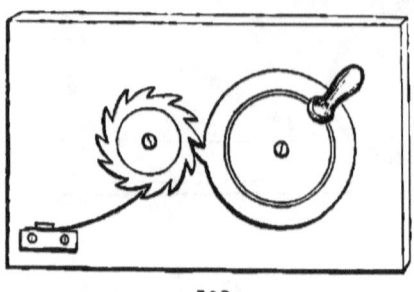

110.

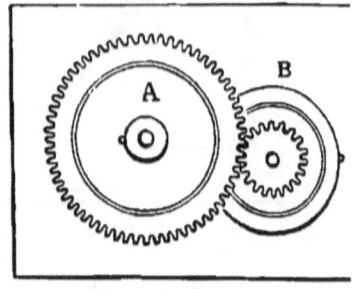

111.

112. *Hooke's Universal Joint.*—Fig. 112. Price 24s.
This joint is formed by a cross, making the diameters of two semi-circles, one of which is fixed at the end of each axis. It is used for the communication of a rotatory motion. It is chiefly used when the inclination is not required to be materially changed; for if the obliquity is great, the rotation is not communicated equably to the new axis at all points of its revolution. The model is of metal, mounted on a polished wooden frame.

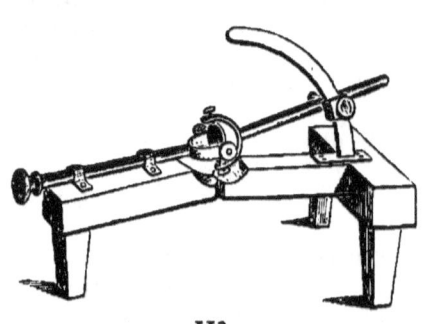

112.

113. *Forge* or *Tilt Hammer*, with anvil and wheel.—The wipers are cut on the edge of the wheel, and a stave fixed in the face of the wheel forms a handle; the whole made of hard stained wood, mounted on a blackboard, measuring 18 inches by 10 inches, with brackets for suspending it to the rail, No. 2. *Price* 7s. 6d.

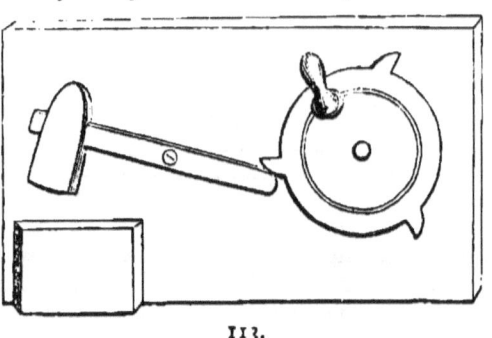

113.

114. *Lock and Key.*—Large wooden model of a lock, showing the spring and tumbler, mounted on a blackboard, measuring 15 inches by 10 inches, with brackets to suspend it from the black rail. Accompanied by an iron key. *Price* 8s.

The figure represents the lock as being set half-way between *locked* and *unlocked*.

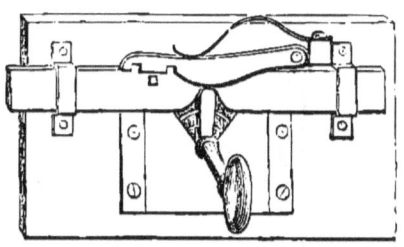

114.

ILLUSTRATIONS OF BRITTLENESS.

115. *Prince Rupert's Drops*, or *Batavian Tears*, for showing the unequal contraction and consequent brittleness of glass that has been suddenly cooled. They are pear-shaped, with a very thin tail; they bear a hard blow on the thick end, but if a small piece is broken off the point, the whole mass flies into dust. *Per dozen, in a box,* 1s.

If the drop is dipped into a thin phial filled with water, and the extreme end is broken off, the sudden concussion breaks the phial.

116. *Bolognian Flask*, for showing the brittleness of glass suddenly cooled, so as to remain unannealed and consequently unequally contracted. The flask is sufficiently strong to bear a smart blow upon the outside; but when a bit of glass or flint, with sharp angles, is dropped into it, the flask cracks. *Per dozen,* 3s. (See *Tyndall on Heat*, page 87.)

117. *Graham's Dialyser*, consisting of a pair of gutta-percha hoops, to strain the dialysing paper, 10 inches diameter; a glass pan to contain water, 13 inches diameter; and two dozen parchment papers. *The set,* 13s.

Other sizes, larger and smaller, are described in *Griffin's* 'Chemical Handicraft,' page 183, where also the process of dialysis is explained.

117.

118. *Graham's Osmometer. Price* 5s. 6d.

119. *Cathetometer*, for use in observing the height of the mercury in barometers, gas-tubes, &c., without approaching so near to the instrument as to change the temperature.—Fig. 119. The stand is not graduated. *Price* 3l. 10s.

The observations are made by a small telescope, that can be fixed at any required height on the cathetometer. The figure represents the form of instrument used by Professor *Bunsen* in his operations of gas analysis.

120. *Model of a Vernier,* in wood, 2 feet long, with scales, for use at lectures. Fig. 120. Price 7s. 6d.

10 lines on the vernier are equal to 11 lines on the scale, so that each division on the vernier is equal to $1\tfrac{1}{10}$ division on the large scale. In making an observation, for example, with a barometer, the zero of the vernier is to be set level with the mercury, the exact height of which on the principal scale is to be measured. There will then be one line of the vernier that coincides with a line on the scale, and the number of that line on the scale indicates the number of tenths to be added to the last whole number on the scale. Thus, if the height to be measured appears to be about 36 and $\tfrac{6}{10}$, the zero of the vernier is to be placed at that level, upon which, looking down the scale of the vernier, it will be found that 6 on its scale is the only line that corresponds with any line of the primary scale.

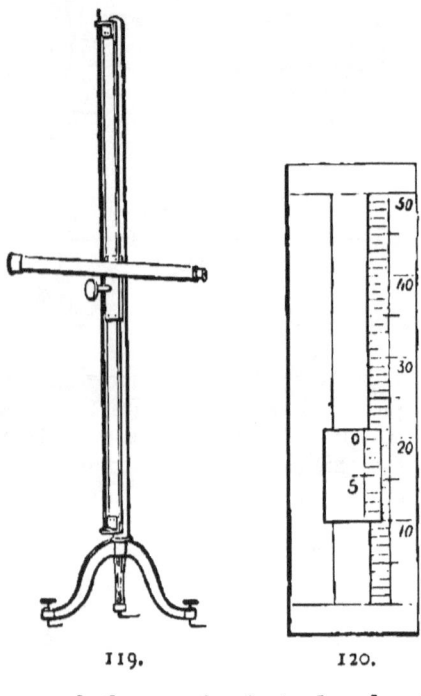

119.   120.

121. *Clinometer,* for measuring the angles of geological strata, brass-bound boxwood rule, 12 inches long, folding up to 6 inches, with spirit level and compass, sunk in the boxwood, and engraved tables. Price 18s.

122. *Balance of large size, for Physical Experiments.*—Length of beam, 2 feet; height from the table, 3 feet; length of

122.

index, 1 foot; with a pair of 5½-inch brass pans with brass chains, and a small pan with hook below, for use in taking specific gravities. Will carry 2 lb. in each pan, and then turn with 1 grain. The upright support unscrews in the middle, for the convenience of packing. Fig. 122. *Price* 31*s.* 6*d.*

This balance is required for class experiments on the specific gravities of solids, liquids, and gases; for the hydrostatic paradox; and for other cases where articles of large bulk have to be weighed.

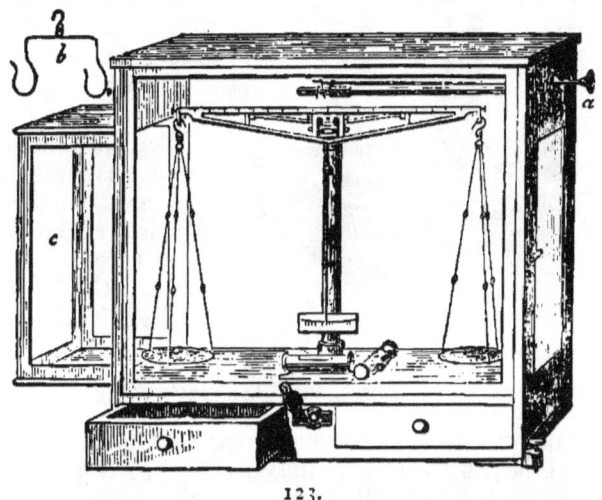

123.

123. *Balance, in glass case, for accurate weighings.*—Beam 15 inches long, with steel bearings, mounted as represented in fig. 123. In mahogany glass case, with a moveable enlargement of the case when large objects are to be weighed. See *c* in the figure; see also *b*, which is a support to attach large objects to the beam in place of the scale-pan: *a* is a contrivance for moving a rider on the beam. The scale-pans are of plated brass, 2½ inches in diameter, and suspended by platinum links. There is a contrivance for steadying the pans (omitted to be shown in the figure). The ends of the beam are formed as represented by fig. 123 A, the pans being suspended by steel hooks resting on steel rings that are sharpened on the inner edge. This balance will carry 100 grammes on each side, and turn with 1 milligramme. It will carry 1600 English grains on each side and turn with $\frac{1}{30}$ grain. When loaded with 500 grains, it will indicate $\frac{1}{100}$ grain. *Price* 11*l.* 11*s.*

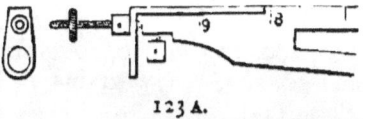

123 A.

D

124. *Balance for the accurate weighing of small quantities.*—Beam, 9½ inches, divided for riders; brass pans, 2½ inches diameter.

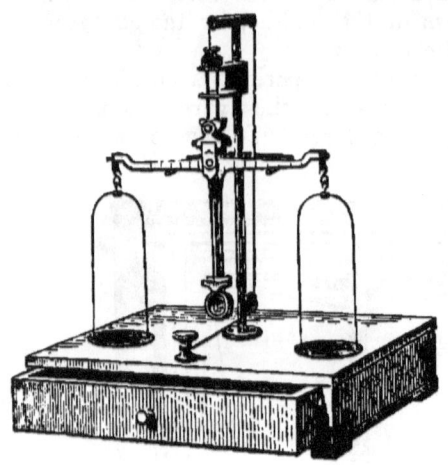

124.

Mahogany box (in which the whole can be packed for travelling), 14 inches long, 7½ inches wide, 4 inches deep. This balance will carry 8 ounces in each pan, and then turn with $\frac{1}{20}$ grain. With 1000 grains in each pan, it turns with $\frac{1}{50}$ grain. *Price* 63s.

For balances of a superior quality and larger size, suitable for use in chemical analyses and for other accurate weighings, consult *Griffin's* 'Chemical Handicraft.'

Commercial Balances.

125. *Scales and Weights*, to weigh quantities up to half an ounce, in plain oak box, with set of grain weights. *Price* 3s. 6d.

126. *Balance* to carry 1 lb. and turn with 1 grain, 10-inch beam, 5-inch brass pans, in oak box. *Price* 18s.

Weights.

127. *Accurate Grain Weights.*—The following set: 600, 300, 200,

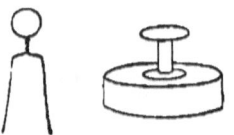

128.   127.

100, 60, 30, 20, 10 grains in brass, fig. 127; and 6·, 3·, 2·, 1·; ·6, ·3, ·2, ·1; ·06, ·03, ·02, ·01 grain in platinum, with riders, fig. 128. In a mahogany box, with tongs. *Price* 28s.

129. *Grain Weights, not quite so accurate as the above.*—The following series: 600, 300, 200, 100, 60, 30, 20 grains, in brass, with knobs, figure 127; and 10, 6, 3, 2, 1, ½ grains, in brass foil, in a divided wooden box. *Price* 12s.

130. *Pound Pile of Avoirdupois Weights*, in brass, round and flat, adjusted, consisting of 1 lb., ½ lb., ¼ lb., 2, 1, and ½ oz. *Per set*, 5s. 6d.

131. *Pound Pile of Avoirdupois Weights*, in cast-iron, round and

flat. Set of 7 weights, adjusted, 1 lb., ½ lb., ¼ lb., 2 oz., 1 oz., ½ oz., ¼ oz.; the two last in brass. *Per set*, 2s.

132. *Single Cast-iron Weights.* 1 lb., 6d.; 2 lb., 1s.; 4 lb., 1s. 6d.; 7 lb., 2s. 6d.; 14 lb., 3s. 6d.; 28 lb., 7s. 6d.
Larger weights can be supplied, if required.

Centigrade Weights.

133. *Accurate Gramme Weights.*—The following series: 50, 20, 10, 10, 5, 2, 1, 1, 1 grammes, in brass; and ·5, ·2, ·1, ·1; ·05, ·02, ·01; ·005, ·002, ·002, ·002, ·001, ·001, in platinum, with riders and lifting-tongs. In a mahogany box. *Price* 35s.

134. *Accurate Gramme Weights.*—A more comprehensive series, containing 500, 200, 100, 100, 50, 20, 10, 10, 5, 2, 1, 1, 1, in brass; and 5, 2, 1, 1, ·05, ·02, ·01, ·01, ·005, ·002, ·001, in platinum. With 4 riders, ivory tongs, and 2 weight-lifters. In a mahogany box. *Price* 52s. 6d.

135. *Gramme Weights*, less accurate than the above described; namely, the series 200, 100, 50, 20, 20, 10; 5, 2, 2, 1; ·5, ·2, ·2, ·1; ·05, ·02, ·02, ·01; in all 18 weights. In a square wooden box. *Price* 12s.

136. *Cast-iron Weights*, French gramme system, from 1 kilogramme to ½ gramme; the series 1000, 500, 200, 100, 50, 20 grammes, in cast-iron; and 10, 5, 1, ½ gramme, in brass. *Price of the set*, 3s.

137. *Standard Weights*, in brass cubes, lettered on all sides. 1000 grammes, 6s.; 500 grammes, 3s. 6d.; 250 grammes, 2s.

Graduated Liquid Measures.

138. *Ounce Measures*, cylindrical form, with foot and spout. Fig. 138.

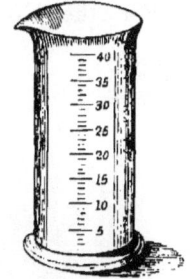

a. 1 ounce, showing spaces of ½ drachm. *Price* 9d.
b. 5 ounces, showing spaces of ¼ ounce. *Price* 1s. 3d.
c. 20 ounces (1 pint), showing spaces of ½ ounce. *Price* 2s. 6d.
d. 40 ounces (1 quart), showing spaces of 1 ounce. *Price* 3s. 6d.

139. *Gramme or Litre Measures*, same form as fig. 138.

a. 30 grammes, showing space of 1 gramme. *Price* 1s.
b. 300 grammes, showing space of 10 grammes. *Price* 2s.

138.

36 MECHANICS.

Gramme Measures, continued:
 c. 600 grammes, showing spaces of 20 grammes. *Price* 3s.
 d. 1200 grammes, showing spaces of 20 grammes. *Price* 4s.
 For a variety of other measures, both English and French, see *Griffin's* 'Chemical Handicraft.'

LINEAR MEASURES.

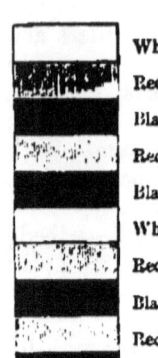

140.

140. *Chromo-Lithographic Linear Scales,* with spaces arranged and coloured, as represented by fig. 140. Width of scale, 1½ inches; vertical depth of each space, 1 inch on the inch scale, and 1 centimetre on the metre scale. On all the scales the colours, beginning at the bottom, are arranged thus: 1, black; 2, red; 3, black; 4, red; 5, white; 6, black; 7, red; 8, black; 9, red; 10, white; and so on to the top of the scale, every fifth space being left white.

141. Prices of the chromo scales, mounted on hard wood, varnished, and with every fifth or white space numbered:
 A. English inch scale, 40 inches long. *Price* 2s.
 B. English inch scale, 20 inches long. *Price* 1s. 3d.
 C. Metre scale, showing 100 centimetres. *Price* 2s. 6d.
 D. Metre scale, showing 50 centimetres. *Price* 1s. 6d.

142. Prices of chromo paper scales, not mounted on wood, and not numbered or varnished, in two slips, rolled up in a small box:
 A. English inch scale, 40 inches. *Price* 4d.
 B. Metre scale, 100 centimetres. *Price* 4d.
 For other chromo-scales, see article *Barometer.*

HYDROSTATICS.

HYDROSTATICS treats of the conditions of equilibrium in liquids and of the pressures which they exert. We shall discuss the heads of the subject in a series of propositions, each accompanied by experimental illustrations.

WATER RISES TO A LEVEL IN COMMUNICATING VESSELS.

200. In communicating vessels that contain a single liquid, the condition of equilibrium is, that the summit of the columns, or the free surface, is everywhere at the same level.

This principle is illustrated by the apparatus, Nos. 201 to 207; and further by Nos. 221 to 224.

201-204. *Water rises to a level in Communicating Vessels.*—Apparatus for this experiment, consisting of bent glass tubes, having two or three upright branches connected with a horizontal branch; each apparatus mounted on a black wooden support. The vessels are nearly all of the same size and style, the wide tube being about 4 inches long and 1 inch in diameter.

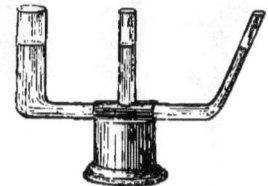

201.

Fig. 201, *price* 2s. 6d.
Fig. 202, *price* 2s.
Fig. 203, *price* 2s.
Fig. 204, *price* 2s.

202.

203.

204.

In experiments on the rise of water in glass tubes, it is convenient to give the water a crimson or blue colour, by dissolving in it a little dye-stuff, such as the following:—

204 A. *Crimson Dye*, for colouring water, to show the level in glass tubes. *Price per bottle, sufficient for colouring several gallons*, 1s.

204 B. *Indigo Blue Dye.* Ditto, ditto. *Price* 1s.

The choice of the colour necessarily depends upon the colour of the backgrounds before which the water tubes are to be placed for inspection.

205. *Water rises to a level in Communicating Vessels.*—Fig. 205. Price 8s.

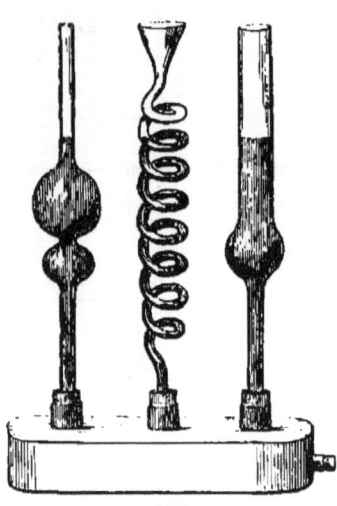

205.

This apparatus consists of a japanned tin cistern of an oval form, with three necks on the upper side, and three tall irregularly-formed glass tubes fitted to its three necks. One of these tubes is surmounted by a funnel. When water is poured into this funnel it fills the oval cistern and then rises to the same level in all the three tubes. The water should be coloured red. After an experiment is made the water may be run off by a small neck in the cistern, which is closed by a cork.

206. *Water rises to a level in Communicating Vessels.*—Fig. 206. Price 6s.

This apparatus consists of a series of five glass vessels, made in three pieces and connected by two ground joints. They are all supported on one foot and can all

206.

be filled with coloured water to the same horizontal level, the water being poured into the middle vase. After an experiment the vessels must be emptied and washed clean before they are set aside.

207. *Water rises to a level in Communicating Vessels.*—Fig. 207. Price 50s.

This apparatus consists of a large bell-shaped vase, mounted on a japanned iron foot, and having a brass-mounted side branch, carrying a moveable vertical tube, with stopcock. It has also six feet of caoutchouc tube, terminated by a glass tube, and having a pinchcock to close the tube. The flexible tube serves to illustrate the conveyance of water by pipes through or across valleys, and to prove the rising of the water through such pipes to the height of the original reservoir.

208. *Spirit Level*, a glass tube, 4 inches long, not mounted in frame. Fig. 208. *Price 6d.*

209. *Spirit Level*, mounted in mahogany, 8 inches long. Fig. 209. *Price 2s. 6d.*

The construction of the spirit level depends on the fact of liquids assuming a horizontal surface. The instrument consists of an hermetically-sealed glass tube, nearly filled with alcohol, fig. 208, and commonly enclosed in a case of brass or mahogany, having the upper side of the tube exposed to view, fig. 209. The tube being very slightly convex in the middle, the instrument is so adjusted that, when it rests on a perfectly horizontal surface, the bubble of air shall occupy the middle point of the tube. Any inclination in a surface on which a spirit level is placed is indicated by a departure of the bubble from the middle point.

207.

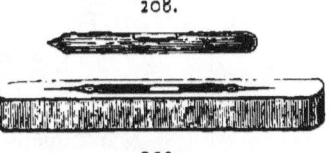

208.

209.

Level of different Liquors in Communicating Vessels.

212. In communicating vessels that contain two different liquids, which do not mix together nor combine chemically with one another, the condition of equilibrium is, that the height of the free surfaces above the level of junction of the two liquids is in the inverse ratio of the densities of the liquids.

This principle is illustrated by the apparatus, Nos. 213 to 217, and yet further by No. 248.

If several separate fluids of different kinds be contained in the same communicating vessels, they will never remain at rest unless all the surfaces intervening between them be horizontal; and this is, in fact, the state of the surface of common liquids which are exposed to the pressure of the atmosphere. See Nos. 215 to 217.

213. *Glass Jar, on foot*, 16 inches high, 2½ inches diameter, with two glass tubes, open and level at both ends, and about 18 inches long. Fig. 213. *Price* 3s. 6d.

Put mercury, to the depth of about an inch, into the jar, and then put in the two glass tubes. The mercury does not rise in the tubes higher than the level in the jar. Then add water to the mercury, *outside the tubes*, till the jar is nearly full. The mercury then rises in the tubes. Measure the height of the mercury contained in the tubes above the level of that outside the tubes, and measure also the height of the water above the level of the mercury in the jar. Compared with one another, the respective heights of the mercury and the water will be as 1 is to 13½, or more exactly 13·6. Now, as the density of mercury is 13½ times that of water, these numbers show that the respective levels of the two liquids above the level of junction are inversely as their specific gravities.

If great exactness in measurement is required, the tubes and jar may be graduated in inches.

214. *Graduated Glass Burette.* Fig. 214. *Price, including the wooden foot*, 4s.

This burette is formed of a wide and a narrow tube, connected at the bottom, and provided with a moveable wooden foot. The wide tube is graduated with a linear scale of 14 parts. To make the experiment, mercury is poured into the tube up to the little mark a. Water is next poured into the wide tube up to the mark $13\frac{6}{10}$ parts. The mercury will then have risen in the narrow tube up to $c = 1$ part, and sunk in the wide tube to 0°. Hence $13\frac{6}{10}$ measures in height of water counterbalance 1 measure of mercury, both liquids being measured from the line of junction at $b\ a$.

215. *Communicating Vessels for two Liquids.* Fig. 215. *Price* 6s.

This apparatus consists of a bent glass tube, each vertical branch of which is about 16 inches long and ¾ inch wide, mounted on a wooden frame with a metal foot and sliding rod, and having a chromo-lithographed scale of inches.

Instructions for use.— 1. Pour into the bent tube as much mercury as will fill the curved base up to the middle of the lowest black mark of 1 inch

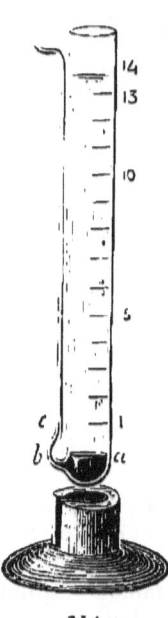

on the chromo scale. Then pour water into the right-hand tube until it rises to the level of $13\frac{6}{10}$ inches on the scale. The water should be tinted blue, to make it visible in a Class. When the water is at $13\frac{6}{10}$ inches in the right-hand tube, the mercury will have fallen in that tube down to 0, and risen in the left-hand tube up to 1 inch, that is to say, to the junction of the black mark with the first red mark on the chromo scale, shown by fig. 215. Consequently you have a column of 1 inch of mercury making equilibrium with a column of $13\frac{6}{10}$ inches of water, these quantities being in the inverse ratio of their specific gravities.

2. Pour water into the left-hand tube until it rises in that tube to the level of $14\frac{1}{10}$ inches by the chromo scale. At that point the water in the right-hand tube will have also risen to $14\frac{1}{10}$ inches, and the mercury will have attained the level of half an inch in both tubes, namely, the same level that it stood at when no water was in either of the tubes. In short, the mercury and water in both tubes are in equilibrium.

3. By means of a long narrow pipette, or a syphon, remove the water from the right-hand tube, without disturbing the mercury or the water in the left-hand tube. When that is done, the mercury will rise to 1 inch in the right-hand tube, and sink to 0 in the left-hand tube, while the water in the left-hand tube will sink down to $13\frac{6}{10}$ inches.

Consequently columns of mercury and water are always in equilibrium when the comparative heights are 1 of mercury to 13·6 of water.

216. *Communicating Vessels for two Liquids.*—This variety of apparatus consists of two tubes of unequal diameters, connected and bent into a U form, each vertical branch being about 16 inches long, mounted on a wooden frame with a sliding support, with a chromo-lithographed scale of inches. Fig. 216. Price 7s.

With this apparatus the three experiments described at § 215 can be performed, and it is needless to repeat the instructions. The following additional experiment is, however, too important to be omitted:—

Put into the bent tube as much mercury as will fill the curved base, and will also

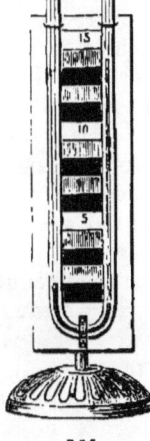

215.

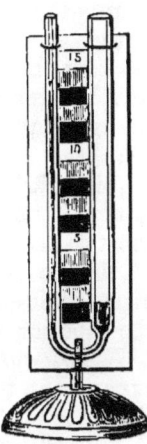

216.

nearly fill the tube up to the 1-inch mark on the chromo scale. The exact quantity wanted is as much as will fill the wide tube up to 1 inch, but will leave the mercury in the narrow tube at 0, when water is added. This necessary quantity can be ascertained by a preliminary trial, and be estimated in cubic inches or otherwise for future use. The mercury being supplied, water is to be poured through a funnel with narrow neck, or added by means of a pipette with slender neck, into the narrow tube until it rises to $13\frac{6}{10}$ inches. If the proper quantity of mercury has been previously supplied, it will then stand at 1 inch in the wide tube and at 0 in the narrow tube. If not, the levels must be adjusted by means of narrow pipettes. When the levels are correct, you have again the result that $13\frac{6}{10}$ inches of water balances 1 inch of mercury, and you have a demonstration of the additional fact that *the diameter of the respective columns in equilibrium is of no importance;* since you find a narrow column of water able to counterbalance a wide column of mercury just as completely as a narrow column of mercury counterbalanced a wide column of water. See 214. These experiments fully justify the principle adduced in § 212.

217. *Communicating Vessels for three Liquids.*—The apparatus represented by fig. 215 can be used for putting three liquids into equilibrium, by the following processes:—

Experiment 1. *Trial of Oil of Vitriol against Water.*—Put into the tube as much mercury as fills the vessel up to the zero line of the chromo scale, that is to say, up to the bottom of the lowest black mark. Put water into the left-hand tube till it rises to the top of the 10-inch mark, then put oil of vitriol into the other tube till it counterpoises the water. The quantity of acid required is 5·4 inches on the vertical scale. But that exact quantity cannot be hit upon at once, excepting after previous calculation. The experimental plan is to add the oil of vitriol a little at a time and watch the level of the two branches of mercury. When the water is first added it destroys the level of the mercury, but the addition of the second liquid restores the level; and, in fact, the experiment is ended when the two liquids are in equilibrium and the mercury in both branches is at the same level. To bring this about, a pipette commonly requires to be used to increase or to diminish the quantity of either liquid, so as to keep the water at 10 inches. The heights of the two columns of liquid are inversely as their specific gravities. To find the specific gravity of oil of vitriol from the above experiment, divide 10 by 5·4, the product is 1·85. After the experiment, the tube can be emptied by means of long pipettes, such as fig. 367.

Experiment 2. *Trial of Sulphuric Ether against Water.*—This experiment is to be made in the same manner as Experiment 1. Begin by putting 10 inches of ether into the tube upon the mercury, and then it will be found that 7·4 inches of water will counterbalance it. Then 7·4 divided by 10·0, gives ·74, which agrees with ether that contains some water. In this experiment, as in No. 1, the sign of the equilibrium of the two liquids is, that the surfaces of the two branches of mercury become level with the lowest black line of the chromo scale, fig. 215.

This experiment cannot be recommended for actual use in determining specific gravities; it is by far too troublesome. See articles on specific gravities in the following pages.

218. *Boyle's Inverted* U *Tube, for estimating the comparative Specific Gravities of two Liquids. Price, graduated, 7s.*

This apparatus is founded on the principle that the heights of columns of liquid which counterbalance one another under the pressure of the air are inversely proportional to the densities of the liquids. *a, b, c,* represent a long and narrow U-shaped tube, having at *c* a neck to which a caoutchouc tube is adjusted. This tube ends with a mouthpiece, *d,* and is provided with a pinchcock for closing it. The long branches of the tube, *a* and *b,* are uniformly graduated from the bottom upwards, and the lower ends dip into two liquids, *e, f,* the respective densities of which are to be tried. On putting the jet *d* into one's mouth and abstracting air from the tube, the two liquids rise in columns, the heights of which are inversely as their specific gravities, and the levels of their surfaces, as at *a* and *b,* can be read on the graduated scales. In this respect the experiment bears a resemblance to those that have been recited between §§ 212 and 217. But the instrument is of little value as a practical method of estimating specific gravities, being much more troublesome to use than many instruments which will be found described in this volume under the head of "Specific Gravity."

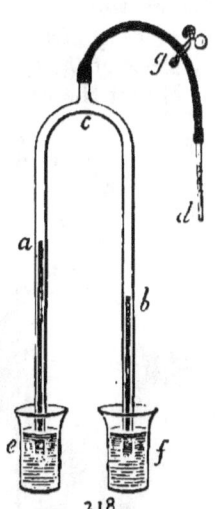
218.

219. *Phial of the four Elements.*—Exhibition of liquids differing greatly in their specific gravities. Stoppered tube, on foot. *Price, unfilled,* 1s.

This toy is intended to contain four liquids that will not combine with, nor decompose, one another. The name recalls the four primordial substances of which the ancient philosophers considered all objects in nature to be composed; but no such significance is now given to the name. The four liquids put into the vessel may be

219.

these: mercury, a solution of salt, alcohol (coloured red), and an essential oil. When this mixture is well shaken in the bottle and then allowed to repose, the mixed substances separate from one another and become superposed in the order of their respective densities, the surfaces of separation being perfectly horizontal.

Pressure of Water in all Directions.

220. Liquids transmit in all directions, and with the same intensity, a pressure exerted at any point of their mass.

This principle is illustrated by the apparatus, Nos. 221 to 224.

221. *Glass Globe, with four necks.*—Size of the globe, 8 to 10 inches in diameter. The tubes, b, c, d, are each about $\frac{1}{2}$-inch or $\frac{5}{8}$-inch wide.

Price of globe and tubes, 25s.
Price of the iron support, 8s. 6d.

Mount the globe, with its fittings, on the iron support, bringing the weight of the globe over the foot of the support.

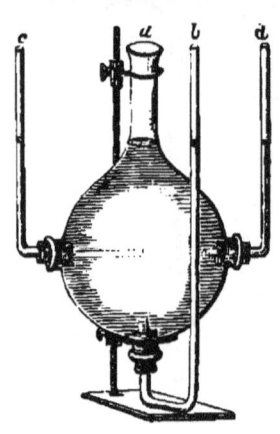
221.

Put a funnel into the neck a, and fill the globe and tubes with coloured water up to the cylindrical part of the neck of the globe, as represented in fig. 221. In this condition the apparatus will illustrate section 200, since it shows that water rises to a level in communicating vessels. But an examination of the apparatus will show that it serves also to illustrate the principle marked § 220 at the head of this section, namely, that liquids transmit in all directions, and with the same intensity, a pressure exerted at any point of their mass. You can, for example, apply the pressure of a weight of water to any of the four necks of the apparatus, and in each case the pressure will be communicated to that part of the mass of water with which the tube is in immediate contact: thus the tube a joins the globe at the upper surface, b joins it at the bottom, c at the centre, and d at one side. From these particular points the pressure is propagated through the whole mass of liquid, and is immediately made sensible at the extremity of each neck, where, after the subsidence of the water-waves, the water attains its level.

When it is necessary to empty the globe, the tube d is turned round in its cork and thus becomes a spout. A piece of caoutchouc

PRESSURE OF WATER IN ALL DIRECTIONS. 45

tube may be attached to it to convey the water in any required direction.

223. *Pressure Apparatus, made of Japanned Tin.*—Fig. 223. Size of the rectangular reservoir, 10 inches long, 5 inches wide, 6 inches high; contents, 300 cubic inches, with six glass tubes. *Price* 35s.

When this apparatus is required for an experiment it must be filled with water, which may be put into any one of the tubes, with the aid of a funnel, until it rises in all the tubes a little above the corks.

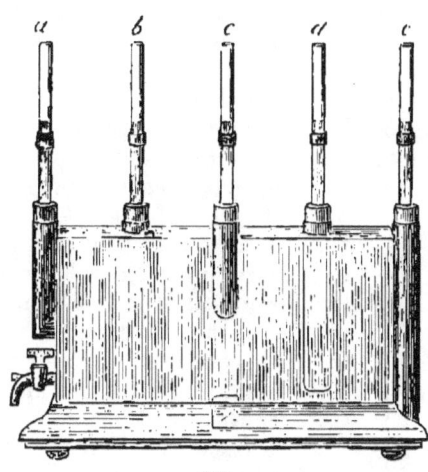

223.

The tubes, a, b, c, d, e, are of glass, about $\frac{1}{2}$ inch or $\frac{5}{8}$ inch in diameter. They are attached to the metal tubes by corks.

The metal tubes are all 1 inch wide. They are inserted into the reservoir as follows:

a opens into the middle of one end of the reservoir.

b opens into the top of the reservoir, but does not dip into it.

c enters the middle of one side of the reservoir, and passes horizontally into the middle of it, so as to act at the centre of the system.

d enters the top of the reservoir, and goes to within an inch of the bottom.

e goes down the outside of the reservoir, passes along the bottom, and enters the reservoir exactly in the middle of the bottom.

All the tubes, both of tin and glass, are open at both ends.

By this arrangement, pressure, by means of a weight of water, can be applied to any part of the system, and it acts first upon the point to which it is applied, and then upon the whole surface of the enclosed liquid and upon the tubes attached to all orifices in the envelope. Thus the tube e applies an upward pressure to the middle of the bottom of the liquid in the reservoir: the tube d applies a downward pressure near the same place; the tube c applies a pressure in the middle of the mass of liquid; the tube b applies it on the upper surface; and the tube a gives a pressure at the middle of one end of the reservoir. From the point directly acted upon, the pressure is immediately diffused through the whole liquid, and is perceptible in the glass tubes even when only a very small quantity of water is used—for example, the tenth part of a cubic inch; and the levels of the liquids in these tubes show the uniformity of the action, and justify the assumption, quoted above, that "liquids transmit, in all directions and with the same intensity, a pressure exerted at any point of their mass."

A stopcock is inserted at one end of the reservoir for the purpose of letting out water to alter the levels in the tubes, or for emptying the apparatus when it is to be put aside. Each glass tube is furnished with a sliding metal ring, to mark the levels of the water. When the apparatus is filled for an experiment, the tubes may be filled to about an inch above the cork connecters, and the rings brought to that level. When the apparatus is in that condition, a very small quantity of water, *put into any tube*, instantly affects the level of the water in the whole set of tubes.

223 A. *Pressure by means of a Piston.*—Fig. 223 A represents a piston formed of a stout glass tube, nearly as large as the bore of the glass tubes *a* to *e*, fig. 223. At the upper end it is closed by a moveable cork or caoutchouc stopper; at the lower end it is tied round with some cotton to make it fit one of the tubes *easily*. To make this act as a piston, the stopper is removed and the piston passed down till the lower end touches the water in the tube; the stopper is then replaced air-tight.

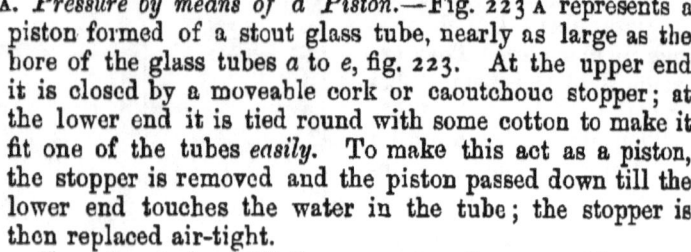

223 A.

Experiment A.—Fill the apparatus, fig. 223, with water, up to about 2 inches from the top of the glass tubes; move the sliding rings to that level. Put the piston, prepared as above described, into one of the glass tubes and press the water down 4 inches, upon which the water will rise 1 inch in each of the other four tubes. This shows that the pressure of a piston applied to water in any tube of this apparatus acts in the same manner as a weight of water will act.

Experiment B. Incompressibility of Water.—The last experiment illustrates to some extent the incompressibility of water; for not the slightest pressure can be exerted by the piston on the water in any one of the tubes without its immediately affecting the level of the water in all the other tubes. No compression or condensation takes place in the body of the apparatus.

Experiment C. Pressure of the Atmosphere.—Fill the apparatus 223 nearly to the top of the glass tubes, and mark the levels. Remove the stopper from the piston 223 A, slowly dip the piston into one of the tubes, and when a depth of 4 inches of water has entered the piston, close the opening at the top with your finger and lift the piston and its contents out of the tube. This abstraction of a column of 4 inches of water from one tube will cause a fall of 1 inch each in the other four tubes, and a little more to restore the level in the

LATERAL PRESSURE OF LIQUIDS.

fifth tube. Of course, the sinking of the water in the four tubes is due to the pressure of the atmosphere; for if the four tubes are previously corked, no such result occurs. This experiment shows the manner in which atmospheric pressure forces water into a common water pump.

224. *Glass Pressure Apparatus.*—Fig. 224. Size of the principal bottle, 6 or 8 pints. *Price, fitted up like fig. 224, 12s. 6d.*

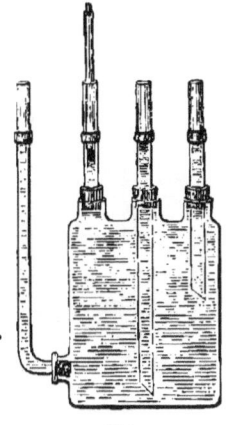

224.

This apparatus consists of a Woulff's bottle of large size, with four necks and glass tubes adjusted by corks and reaching to various parts of the interior of the bottle. Thus one tube goes halfway down one side of the bottle, another goes to the centre of the bottom, a third merely goes into the top of the bottle, and the fourth enters the bottle at the bottom on the outside. In one tube there is a piston similar to that marked No. 223 A, and serving for similar uses. After the detailed descriptions given of the apparatus from No. 221 to 223 A, it seems to be only necessary here to say that this apparatus is, like the others just referred to, intended to prove the truth of the law laid down in § 220, that liquids transmit in all directions, and with the same intensity, a pressure exerted at any point of their mass.

LATERAL PRESSURE OF LIQUIDS.

227. *Jar for showing the Lateral Pressure of Liquids.*—Fig. 227, about 6 inches high and ⅝ inch wide, the lower orifice fitted with a cork and narrow tube. *Price 1s. 6d.*

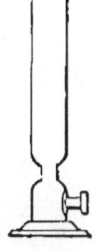

227.

The apparatus, filled with water, is placed on a large flat cork or a light porcelain capsule, on a surface of water in a large pan. When the stopper is removed from the end of the narrow tube, and the water spirts out, the apparatus swims off in the opposite direction.

228. *Barker's Mill* (fig. 228).—The moving cistern is made of japanned tinplate, 8 inches high, 2 inches wide, with 4 sprays; mounted on a frame; so constructed that the action can be seen at a distance. *Price 7s.*

The mill can be kept in action by a supply of water poured in at the top of the cistern. The round disk at the top of fig. 228 is

HYDROSTATICS.

a cork, put to show where driving-bands for mill-work may be applied to this mill.

229. *Tourniquet Hydraulique* (fig. 229).—A variety of Barker's Mill, a conical pear-shaped glass vessel; mounted with brass fittings and polished black wood stand. Height 2 feet, trough 14 inches in diameter, base 17 inches square. *Price* 2*l.* 10*s.*

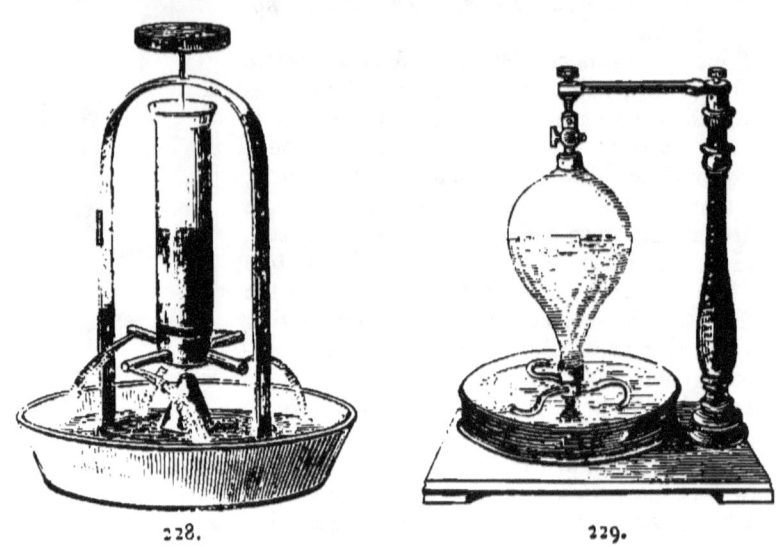

228.   229.

This apparatus is mounted so as to move easily on a vertical axis, which carries a stopcock on the upper part, and at the bottom 3 very narrow and bent horizontal tubes, each giving 3 minute jets of water. To fill this apparatus, it is unscrewed at the stopcock and water is poured in by a funnel. It is then screwed up and the stopcock turned off, upon which the apparatus remains out of action, the atmospheric pressure being cut off. But when the stopcock is opened the water flows out at all the jets, and the glass vessel revolves in a direction opposite to that of the jets of water.

UPWARD PRESSURE OF WATER.

230. Water presses upwards as well as sideways and downwards. This fact is proved by Experiments 231 to 234.

231. *Upward Pressure of Water.*—The apparatus to show this fact consists of a glass cylinder open at both ends; a disk with a cord to close the bottom of the tube; and a jar to contain water. Fig. 231. *Price of the set,* 4*s.* 6*d.*

232. Apply the disk to the cylinder so as to close it water-tight, and then plunge the cylinder into the water contained in the jar. At a certain depth the string may be let loose, because the pressure of the water from below against the disk will prevent the entrance of water into the cylinder. Raise up the cylinder gradually. At a certain point near the surface of the water the disk will fall off, because the pressure of the shortened column of water is become insufficient to support the weight of the disk.

233. Or, instead of raising the cylinder, you may pour into it *with precaution* (as by a bent funnel) a quantity of coloured water. You will thus gradually reach a point where the upward pressure of the water on the jar is overpowered by the downward pressure of the coloured water poured into the cylinder. The disk then falls off.

234. *Upward Pressure of Water* (fig. 234). *Another Experiment.*—Apparatus consists of a glass tube, with a piece of flexible caoutchouc, or of thin bladder, tied over one end to form a bag. *Price of* 234 a, 1s. 6d. 234 b, glass jar, *price* 1s.

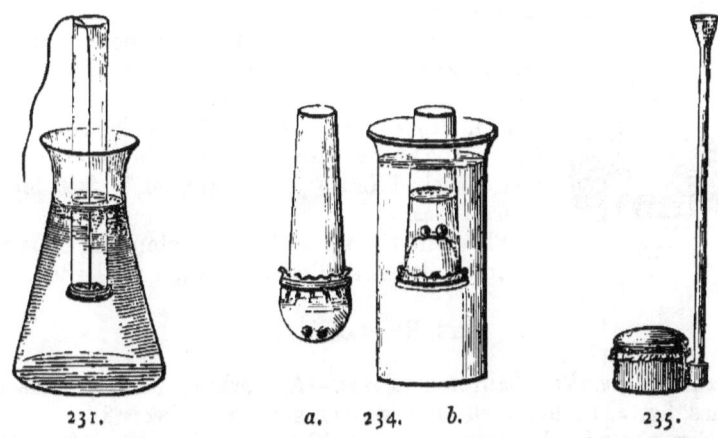

231.   a. 234. b.   235.

Fig. 234 a represents the apparatus out of water, b the apparatus placed in water contained in a beaker. A couple of marbles or stone balls, ½ inch in diameter, are represented in the figure. A little water, coloured blue or red, put into the tube, helps to make the action visible at a distance.

Bladders used for such experiments can be preserved in good condition by being washed with a solution of carbolic acid or a mixture of glycerine and water.

235. *Hydrostatic Bellows, to show the Upward Pressure of Water, in Communicating Vessels.*—This apparatus consists of a cylinder of japanned tinplate measuring 7 inches in diameter and 2½ inches in

E

height, closed by a stout sheet of india-rubber. To the side of it, near the bottom, is soldered a metal arm, which carries a long funnel 36 inches high, and $\frac{5}{8}$ inch in diameter. Fig. 235. *Price* 8s.

When this apparatus is to be used the lower part must be filled with water and the rubber be tied over, with exclusion of air, the performance of which process requires some care. When the air is all replaced by water, and the caoutchouc cover is tied firmly on, water may be poured into the long tube. The caoutchouc then rises and is able to push up a considerable weight, a board being laid on the caoutchouc and the weight on the board. The elevating power of this apparatus is that of a column of water, as wide as the large pan, and of the height of the column of water in the narrow tube. But the material of which the apparatus is constructed is not sufficiently strong to sustain so great a pressure. Hence, when large weights are to be raised, the more powerful Hydrostatic Bellows must be used.

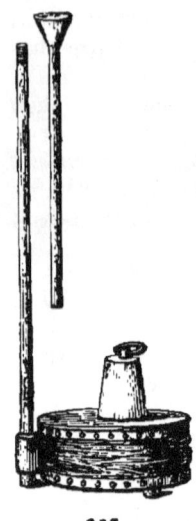

237.

236. *Hydrostatic Bellows*, 12-inch circular black boards, with a vertical jointed funnel or tube, 6 feet high. *Price* 24s.

237. *Hydrostatic Bellows*, of superior construction to the above, with polished mahogany boards, and polished brass jointed funnel, 6 feet high. *Price* 3l. 3s.

This will rise up with a man standing upon it, when the tube is filled with water to the top.

THE BRAMAH PRESS.

241. *Bramah's Hydrostatic Press.*—A working model, made in glass, fig. 241; mounted on a wooden support. *Price* 8s.

Fig. 241 *a* is the force pump. When the piston is lifted, the valve below it rises and lets in the water. When the piston is forced down, the valve below it closes, and the valve in *b* rises and lets in the water to the cistern *c*. This process being repeated, the large plunger *d* is soon acted upon, and though the apparatus is small and fragile, it has power to lift a considerable weight put on the flat summit of the plunger *d*. Between the cistern *c* and the plunger *d*, there is a stuffing or ring of cotton.

242. *Bramah's Hydrostatic Press.*—Working model of, in brass, the table of an oval form, size 5 by 2¼ inches; diameter of small plunger, $\frac{1}{8}$ inch; diameter of large plunger, ¾ inch. Fig. 242. *Price* 4l. 14s. 6d.

THE BRAMAH PRESS. 51

242 A. A variety of this press with a vertical pump, acted on by a lever, and easy to work. *Price 6l.*

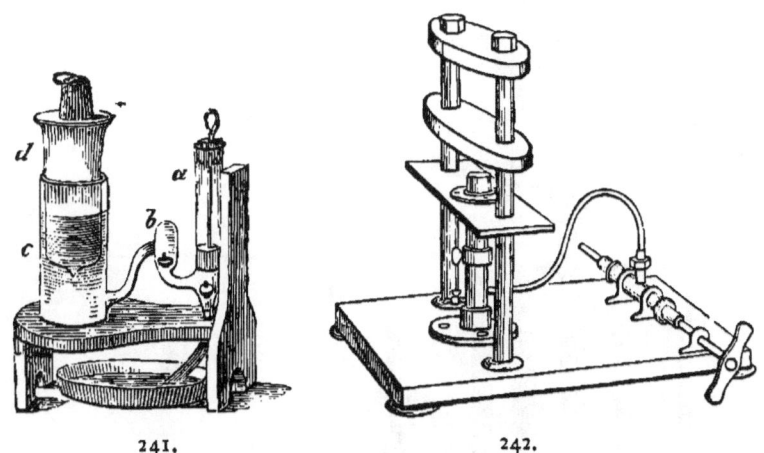

241.   242.

243. *Bramah's Hydrostatic Press.*—A working model. Size of plates for pressing flat objects 5 by 5 inches, with 3 moveable knife edges for breaking bars of metal, and a circular table 12 inches diameter for raising weights; mounted on a japanned iron box, 15 inches long, 10 inches wide, very powerful, both for pressing and lifting. Fig. 243. Diameter of the small plunger, $\frac{1}{4}$ inch; diameter of the large plunger, $1\frac{5}{8}$ inch; with a safety valve. *Price 15l.*

Letter *a* is the small pump for forcing water into the apparatus, from a cistern contained in the base; *b* is the cylinder of the large pump; *c* is the upper and *d* the lower plate of the flat press; *e* is the flat table upon which objects are placed to be lifted. It was by means of apparatus of this character that the large and heavy portions of the Britannia Tubular Bridge were raised into their proper positions; *f* and *g* are two cutting edges that fit into holes in the plate *d*, and *h* is a similar cutting edge that can be fitted into the lower face of the plate *c*, after removal of the table *e*. A bar of metal then placed between the cutting edges as shown by fig. *f*, *g*, *h*, can be cut in two. Fig. *i* is a block for closing a hole in the plate *c* when the table is removed. The stopcock represented in the front of the figure, connected with a curved pipe, is intended to remove the pressure and let the water out of the press.

244. *Calculation of the power of the Hydrostatic Press.*—Suppose the piston of the small pump to be $\frac{1}{2}$ inch in diameter, and that of

the large cylinder to be 5 inches. Then the ratio of their areas will be as 1 to 100; and suppose the piston to be worked by a lever gives an advantage of 5 to 1, then, for every pressure of 1 lb. put on the lever, a pressure of 500 lb. will be exerted by the press.

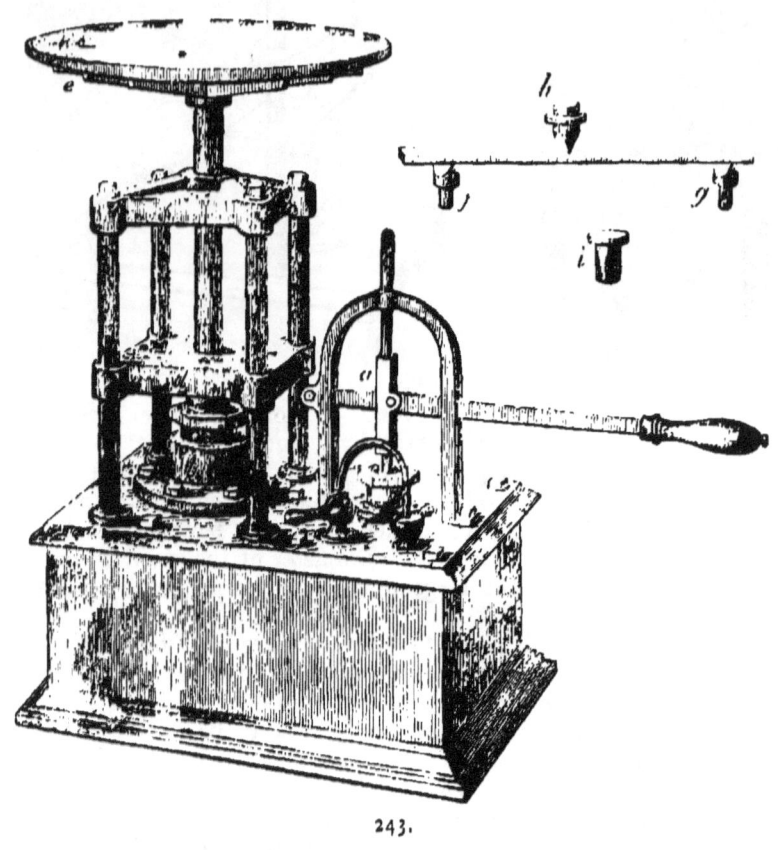

243.

Hydrostatic Paradox.

245. The pressure exercised by a liquid upon the horizontal base of the vessel that contains it, is equal to the weight of a cylindrical column of that liquid having for base, the dimensions of the bottom of the containing vessel, and for height, the distance from the bottom to the plane of the free surface. *Pascal.*

246. This principle of Pascal's is held to be sufficiently proven by the experiments that are performed by means of the apparatus

HYDROSTATIC PARADOX. 53

known as *Pascal's Vases* and *Haldat's Apparatus*, Nos. 247 and 248.

247. *Pascal's Apparatus, modified by M. Masson.*—Fig. 247. Price, *without the balance*, 2l. 2s.

" We owe to Pascal the discovery of the Hydrostatic principle, to which his name remains attached, namely : That the pressure exercised by a liquid on the bottom of a vessel depends exclusively on the dimensions of that bottom and on the height of the column of liquid which it supports. Various apparatus have been constructed for the experimental demonstration of this principle. I shall describe that which *Pascal* himself imagined, and which has been brought to perfection by *M. Masson.*

" Three glass vases of different forms, fig. 247, open at both ends, can be screwed by brass mounts attached to the lower end of each of them, upon a brass collar, which is fixed upon a tripod. The lower orifices of these three vases, which are all of exactly the same diameter and are carefully ground, can be closed one at a time, by means of a disk, which disk is held by a thread, that traverses the vase and is attached to the arm of a balance. Weights which are put into the scale pan attached to the other arm of the balance, apply the disk exactly against the lower orifice of the collar into which the vase is screwed.

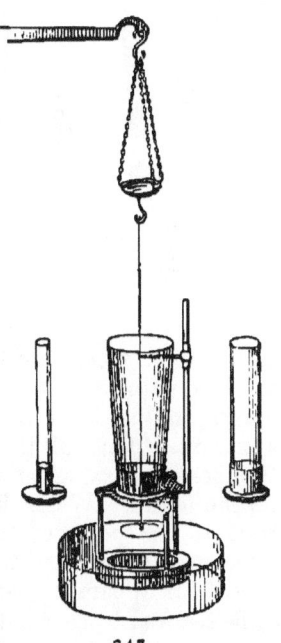

247.

" For example, one of the vases, being screwed on the brass collar, we pour into it a certain quantity of water, so much, indeed, that its pressure on the disk makes a just equilibrium with the weights that have been put into the opposite scale pan of the balance. The level to which the water rises in the tube is marked by the indicator—a pointed rod which slides on an upright rod which rises from the tripod. We remove the weights from the balances, and, the equilibrium being destroyed, the disk is repelled from the vase, the water escapes into the glass pan placed below it, and the vase is emptied.

" The vase is then unscrewed, and is replaced by another vase. We put again into the balance pan the weights that had been removed, and the second vase being thus closed by the disk, we again put water into it, and we ascertain that the height of the column of water capable of bringing into equilibrium the weights on the balance is exactly equal in the two vases, although their forms and capacities are different.

"We repeat the experiment with the third vase, and find that the result is the same, not only with it, but with any other vase the lower orifice of which has the same diameter."—*Salleron*.

248. *Haldat's Apparatus*, for showing that the pressure of liquids depends upon the height and the extent of surface of the bottom of the columns, and not upon the capacities of the vessels. The apparatus is represented by fig. 248. It consists of three large glass vases, mounted with brass collars and screws of uniform size; it has a bent horizontal iron tube terminating at one end in an iron cup, to which the glass vases can be screwed one at a time, and at the other end with a vertical glass tube upon which slides a small brass ring. There is an upright brass rod with a moveable point to mark the surface of the liquor in the jar that is in operation. *Price of the set*, 2l. 2s.

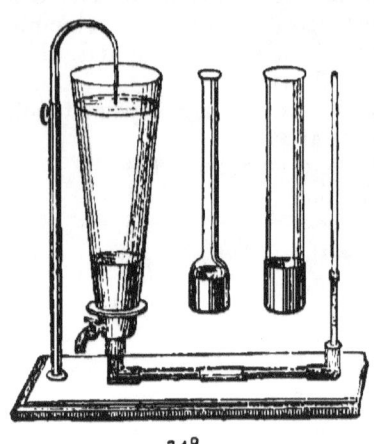

248.

249. When the apparatus is to be used for an experiment the horizontal tube must be filled with mercury, which must be poured into the cup where the stopcock is fixed, but not so much of it as to cause it to run into the stopcock. The mercury then rises to a corresponding level in the narrow tube at the other end of the apparatus. One of the glass jars is then to be screwed on to the mercury box, and the brass pointer is to be adjusted, as shown in the figure. Water is then to be poured into the jar until it rises to the point of the indicator. When this is observed, the mercury will be found to have risen in the narrow tube at the other end of the apparatus; and the exact point to which it has risen is to be marked by means of the sliding ring. The water is then to be run off by the stopcock. Meanwhile the mercury in the narrow tube sinks down to its original level; but the position of the sliding ring is not to be altered. The glass jar is next to be removed and exchanged for another, which in its turn is to be filled with water till it rises to touch the point of the indicator. It will then be seen that the mercury has again risen in the narrow tube up to the sliding ring. The operation is to be repeated with the third glass jar, and it will be attended with the same result,

namely, that the three vessels when filled with water up to the brass pointer, all drive the mercury in the narrow tube up to the same level.

Note on Haldat's Apparatus.—In most books on Physics this apparatus is represented as sharing with Pascal's vases the property of demonstrating the truth of Pascal's principle, laid down in paragraph 245. We think it fails to give any such demonstration, and that it rather belongs to the apparatus described as illustrating paragraph 212, which treats of the level assumed by two different liquids in communicating vessels. The proof of this is, that for every inch which the mercury rises in the narrow tube of Haldat's Apparatus, the water must rise 13·6 inches in the large tube, whatever may be the form and capacity of the wide tubes, and whatever the area of their bases. The exact correspondence of the diameters of the bases of Haldat's vases does not *hinder* the performance of his experiments, and does not *help* it. Read paragraphs 212 to 216.

APPARATUS FOR DETERMINING THE PRESSURE ON THE BASE OF A CONE OF WATER IN A VESSEL WITHOUT A FIXED BOTTOM.

250. *Cone.*—Fig. 250 represents the vessel employed for this purpose. It is a conical glass tube, 10 inches long, of 2 inches interior diameter at one end, and 2¼ inches at the other end; or very nearly of these dimensions. Each end is made as level as possible, has a broad welt, and is finely ground. *Price 3s. 6d.*

251. *Cone mounted for use.*—Fig. 251 represents the glass cone mounted for use. It is supported by an iron collar, attached to a strong iron support. The cone is protected by slips of caoutchouc tied on under the iron collar, without which it cannot with safety be screwed up tight. A ground disk of plate glass, 3 inches in diameter, is supported by a thread to the hook of a balance pan, namely, the small pan represented in fig. 122, which shows the balance that is suitable for

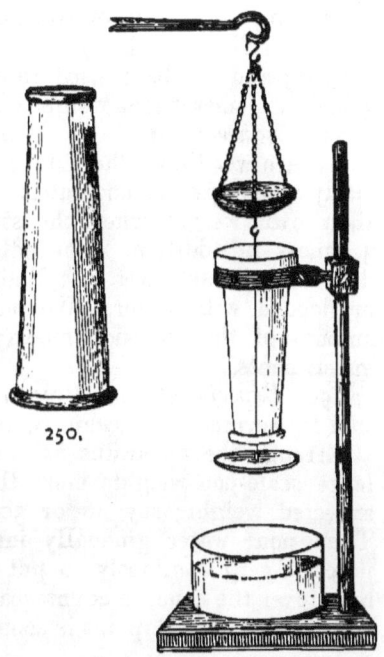

250.

251.

these experiments. To get this thread up through the cone, a long iron wire with a hook at the end is used. A glass pan is placed on the foot of the support, ready to catch the water that, from time to time, escapes from the cone.

When the glass cone is properly fitted to the collar of the iron support, it can be inverted without derangement. The whole system is perfectly rigid, and does not give way when the cone is filled with water. *Price of the Apparatus represented by fig. 251, without the balance*, 1l. 5s.

252. *Experiments with a Glass Cone*, 10 inches long, 2 inches internal diameter at the narrow end, and 2½ inches at the wide end.

253. *Experiment A. Determination of the capacity of the Cone in Ounces of Water.*—Fit up the apparatus as shown by fig. 251. Put into the large scale-pan of the balance enough weights to draw up the disk firmly against the bottom of the glass cone. Then pour water slowly into the cone from a quart glass measure graduated into ounces, such as fig. 138 d. Fill the cone level with the ground welt, and then examine what is left in the graduated measure. In an experiment made thus with a cone of the above dimensions, the quantity of water required to fill it was 24½ ounces.

254. *Experiment B. Determination of the Weight of the Disk.*— The disk and string, in a dry state, separate from the rest of the apparatus, can be weighed in a balance, and the weight be used in estimating the results of some of the following experiments. We have used disks of thin ground glass, of 1 ounce and 1½ ounces in weight; and we have used thick disks of ground plate glass of 3 inches diameter that weighed 6 ounces.

255. *Adhesion of the Disks to the Cones.*—When the disks and cones are wet they adhere to one another with some force, especially when ground accurately. Not only will the disks support their own weight when the string is slackened, but often 2 or 3 ounces in addition. The adhesive power is greatest when the disk is moderately wet. It diminishes when the disk is either dry, or flooded with water. We have found no way to estimate the amount of the adhesion modifying the results of the following experiments.

256. *Experiment C. Pressure on the base of the Cone, placed with the narrow end downwards, and weighed on the loose disk.*

Arrange the apparatus as shown by fig. 251, and put into the large scale-pan weights more than enough to counterbalance any expected weight; say 28 or 30 ounces for the water and disk. Then pour water gradually into the cone till it is full. Next proceed very cautiously to put weights into the small scale-pan hung over the cone, to counterbalance the excess of weight in the large scale-pan. A point is soon reached, when the disk gives way and lets out a little water. Take out the weight last added, fill up the cone with water, and add a smaller weight. By one or two changes you finally reach the point where the water is exactly counterbalanced. You may then run out the water and examine the weights in the pans, deducting from the gross weights in the large pan, first the counterpoise of the disk, and then the weights in the small pan; the residue shows the pressure on the base of the water in the cone. In an experiment with this cone the pressure of water thus determined was $23\frac{1}{2}$ ounces.

257. *Experiment D. Pressure on the base of the Cone placed with the wide end downwards, and weighed on the loose disk.*

Arrange the apparatus as represented by fig. 251, but turn the cone with the wide end downwards. Put 40 ounces of weight into the large scale-pan, and proceed as directed under Experiment C, § 256. The pressure of the water, determined in an experiment made thus with the cone, was 34 ounces.

The results of the foregoing three experiments are these:—
A. Quantity of water that fills the cone, $24\frac{1}{2}$ ounces.
C. Weight on the disk, narrow end downwards, $23\frac{1}{2}$ ounces.
D. Weight on the disk, wide end downwards, 34 ounces.

The vessel, the water, the balance, the weights, were the same in all cases.

These facts seem to constitute a *Hydrostatic Paradox*, though that term is commonly applied to the Hydrostatic Bellows, described at §§ 235 to 237.

Specific Gravity.

EQUILIBRIUM OF SOLIDS WHEN IMMERSED IN LIQUIDS.

Principle of Archimedes.

270. Every solid when immersed in a liquid loses a portion of its weight equal to the weight of the liquid which it displaces.—*Archimedes.*

271. The following three cases can occur on putting this principle into practice, in reference to water taken as a standard:—

A. The solid may be heavier than its own bulk of water. In that case it will sink.

B. The weight of the solid may be equal to that of the water it displaces. In that case the solid will rest indifferently in any part of the water.

C. The weight of the solid may be less than that of its own bulk of water. In that case it swims on the surface of the water; and its weight is equal to that of the water which is displaced by its submerged portion.

These three cases of differences in density are illustrated by experiments 272 to 274.

272. *Exhibition of Differences in the Specific Gravity of Liquids.*
 a. Half fill the jar a, fig. 272, with a strong solution of common salt, and put an egg into it.

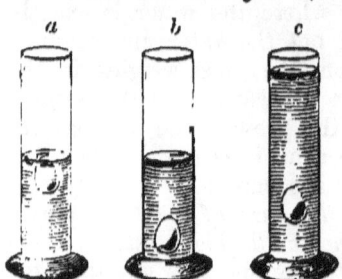
272.

 b. Put an egg into the jar b, and half fill the jar with water free from salt.

The egg will swim in the salt water, but sink in the fresh water, because its specific gravity is a little greater than that of the fresh water, and a little less than that of the salt water.

 c. By cautiously adding fresh water to salt water a mixture can be made in which an egg will float, if it is gently put in above, and allowed to sink till it finds a stratum of liquid of its own density. This it will do after a little oscillation. Fig. c.

Price of the Jars, fig. 272, 12 inches high, 3 inches wide, 1s. 8d. each.

273. *The Bottle Imp.*—This is a slight glass figure, which can be readily changed in specific gravity, so that it will swim in a jar at the top of the water, in the middle, or at the bottom, as desired. It serves, therefore, to illustrate the above cited three cases of differences in density. See *Bottle Imps,* § 730, among the Pneumatic Experiments, for a full explanation of this toy.

274. *Three Solids of Different Densities.*—Into a glass jar, 12 inches high and 3 inches wide, put a saturated solution of common salt to the depth of 6 inches. Into this solution put a ball of ebony, an egg, and a ball of beech-wood. The ebony ball sinks to the bottom, while the beech ball and the egg swim. Now add water gradually and stir the mixture with a long glass rod. At a particular stage of dilution the ebony ball will remain at the bottom, the beech-wood ball will rise to the surface, while the egg will remain suspended midway in the liquid. *Price of the two ebony and beech-wood balls,* 1s. *Price of a glass jar 10 inches high and about 3 inches wide, and glass rod,* 2s.

274 A. If a handful of ebony and beech-wood balls of the same size are dropped into water contained in a glass beaker, the ebony balls instantly go to the bottom and the beech balls rise to the surface of the water. The separation of the balls is so sudden and so complete, that, simple as the matter is, the effect is surprising.

SPECIFIC GRAVITY.

275. Experimental Demonstration of the Principle of Archimedes.—The apparatus consists of A, a cylinder of polished brass closed at both ends and having a hook at the top, its weight being 1000 grains; and B, a cylinder of brass open at the top, of such capacity as exactly to contain the brass box A, and having a ring handle. In fig. 275 the two cylinders are represented as connected with the beam of the Hydrostatic Balance, No. 122, with the closed box sunk into water. *Price of the pair of brass cylinders,* 6s.

276. *Experiment A.*—Attach the two cylinders to the small pan as represented in fig. 275; but without using the water jar. Counterpoise the cylinders by weights put into the opposite scale-pan of the balance.

Experiment B.—Bring under the cylinders a jar containing water, and raise the jar until the closed cylinder is just immersed in the water. A short water jar (fig. 275 B, 1s.), and a mahogany table support (fig. 275 b, 3s. 6d.) below it, is more convenient than the tall jar shown in the figure, because the arrangement permits the raising of the water without disturbing the balance. As soon as the brass closed box is immersed in the water the equipoise is destroyed, the small scale-pan rises and the large pan with the weights sinks.

Experiment C.—If then the open brass cylinder is filled with water, applied slowly by means of a pipette till it is full, the equilibrium will be restored. The upper part of the brass box will be level with the surface of the water; the balance beam will be horizontal and the needle vertical; and the whole will be at rest; the upward pressure against the brass box in the water being exactly equal to the downward pressure of the water put into the hollow cylinder, that is to say, to the weight of the water which the box had displaced. The closed box weighs 1000 grains; it has the bulk of 1000 grains of water, and the capacity of the hollow cylinder is that of 1000 grains of water.

Admitting that the above described experiments clearly demonstrate the truth of the principle of Archimedes, how are we to explain the following experiment?

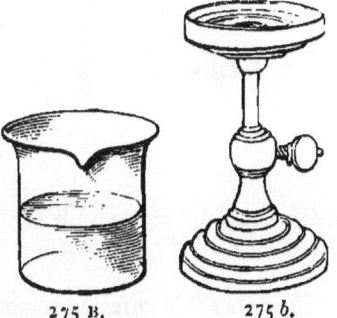

275.

275 B.   275 b.

HYDROSTATICS.

277. A glass jar containing water and a brass closed cylinder, such as is described above, are placed side by side on the scale-pan of a balance and are counterpoised by weights put into the opposite scale-pan. The brass cylinder is then taken up and put into the water contained in the glass jar. But this immersion fails to destroy the equilibrium that was previously established. We have proved above that the brass box loses 1000 grains in weight when immersed in water. Why is the immersion in this case without effect?

Is it not evident that what the box loses in weight is gained by the glass and water, which acquire the power of a downward pressure to compensate for the upward pressure that existed against the box? This can be verified by an experiment made with the apparatus represented by fig. 278.

278. Weigh a glass beaker partly filled with water; then immerse in the water the closed brass cylinder supported externally by the method shown by fig. 278. When the brass cylinder enters the water the equilibrium is overset, and the balance beam descends on the side of the beaker. By how much does the immersion of the brass cylinder, though supported externally, augment the weight of the water? Precisely by the weight of the water displaced, namely, 1000 grains. To prove this fact it is only necessary to remove, by means of a pipette, as much water from the glass beaker as will fill the suspended hollow brass cylinder. When the latter is thus filled, the equilibrium will be restored.

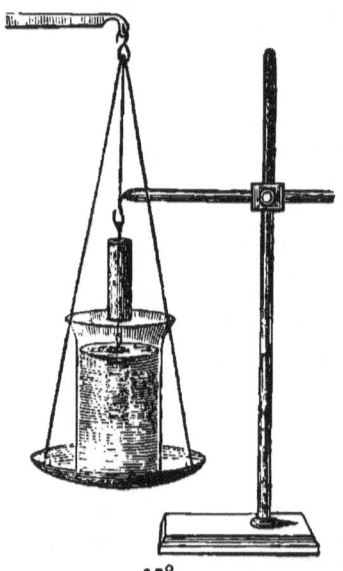
278.

The theoretical explanation of the above facts is very simple. When we plunge a solid body into water we raise the level, and consequently we augment the pressures in the vessel to the same extent as we should do by adding a volume of water equal to that which the immersed solid displaced.

279. *Price of the Stand for supporting the brass cylinders, fig. 278,* 10s. 6d.

This stand is similar to the support of No. 251, excepting that the vice is removed from the block, and a rod with a hook at the end substituted for it.

Exhibition of various methods of Estimating the Specific Gravity both of Solids and Liquids.

280. The principle of Archimedes (No. 270) serves to resolve a very interesting problem, that, namely, of estimating the comparative densities, or specific gravities, both of solid and liquid bodies. We shall describe briefly some of the principal methods of effecting the solution of such problems.

In all the specific gravity estimations, both of liquids and solids, the standard taken for comparison is *pure Water* at 62° of temperature by Fahrenheit, the barometer being at 30 inches, and both water and other substances supposed to be weighed in the open air against weights made of brass. The only exception is alcohol, which, by Act of Parliament, is valued at 60° Fahrenheit. When French weights and measures are referred to, the conditions of the French standards are assumed to be adhered to.

281. *Estimation of the Specific Gravity of a Solid Body by means of the Hydrostatic Balance.*—First, the solid is to be weighed in the air. Which of the balances described between Nos. 122 and 125 should be used for this purpose depends upon the bulk or weight of the solid, and upon the degree of accuracy to be aimed at in the operation. We shall refer to the Hydrostatic Balance, No. 122, which has been already several times referred to, as suitable for Class Experiments in Hydrostatics, but with the notice that it is unfit for working with very small quantities. Read the descriptions of Nos. 122 to 137.

The weight of the object being found, the next step is to determine the weight of an equal volume of water. *The division of the former weight by the latter gives the required density.* The determination of the weight of an equal volume of water may be made as follows:—

282. The solid body under trial may be suspended by a thread, a hair, or a fine wire, to the small pan of the hydrostatic balance. It is to be counterpoised in air, and is afterwards to be plunged into water, which destroys the equilibrium. The weight which it loses is precisely equal to that of the water which it displaces, and is exactly that weight of a volume of water equal to the volume of the solid which we wish to ascertain. To prove this fact it suffices to place the indicated weight upon the small pan to which the submerged object is suspended, which immediately restores the equilibrium.

Example *a*.—A solid weighed 400 grains in air. In water it weighed 300 grains. Hence the weight of its volume of water was 100 grains. Then, $\frac{400}{100} = 4$; so that the specific gravity of this solid was 4000, water being 1000.

Example *b.*—A piece of copper weighed 250·70 grains in air. In water it lost 28·48 grains. Then, $\frac{250\cdot70}{28\cdot48} = 8\cdot80$, which is the specific gravity of the piece of copper, water being 1·00.

283. *Estimation of the Specific Gravity of a Liquid by means of the Hydrostatic Balance.*—Weigh a solid body first in air, and then in water, and also in the liquid whose specific gravity is to be estimated. The loss in each of the last two cases will be that of the weight of a volume of each liquid corresponding to the bulk of the solid submitted to trial. The solid to be weighed ought to be a globe of glass about $\frac{3}{4}$ inch in diameter, containing a little mercury, and it should be fastened to the hook under the small pan of the hydrostatic balance by a thread or fine wire. When first weighed in air, and then in water, the loss shows the weight of a volume of water equal to the volume of the solid. Then having carefully cleaned and dried the solid, and weighed it in the liquor to be assayed, the second loss gives the weight of an equal volume of that liquid. Dividing the last result by the former, we have for quotient the required density.

Example.—In an experiment made with alcohol the loss of weight was 354·3 grains, and with water the loss was 442·8 grains. Then $\frac{354\cdot3}{442\cdot8} = 0\cdot8$, which was the specific gravity of the alcohol submitted to trial.

Instead of a spherical glass weight a species of cylinder, or plummet, may be used with advantage, especially as this form of solid can be provided with a small thermometer for determining the temperature of the liquids submitted to experiment. See *Mohr's Specific Gravity Balance*, No. 299.

284. *Estimation of the Specific Gravity of a Solid Body by weighing it in a Bottle.*—The bottles for this purpose are represented by *a* and *b*, fig. 284; they are made of slight blown glass, with wide mouths and hollow stoppers, capacity about half an ounce of water, not adjusted.

Method of use.—Fill one of these bottles with distilled water up to the point of the stopper, and place it on the scale-pan of a balance, and by the side of it place some fragments of the body to be tried, which must be small enough to enter the mouth of the bottle, but not be in powder, say 100 grains accurately weighed. One of the finer balances, Nos. 123, 124, should be used for this experiment. Counterpoise by weights put into the opposite scale-pan.

a 284. *b*

Then put the solid body into the bottle, from which, of course,

SPECIFIC GRAVITY.

there issues a volume of water equal to the volume of the solid; put the stopper into the bottle and see that it is as full of water as it was at first; wipe the bottle dry on the outside, and put it again on the balance pan. It will not produce equilibrium, but will require a compensation for the loss of the water that was displaced, say 25 grains. The density of the solid in that case will be $\frac{100}{25} = 4$. Here, as in the two former cases, the specific gravity of the solid body is the expression of *the number of times which it is heavier than an equal bulk of water*; and that expression is found by dividing the weight of the solid in air by the loss of weight which it suffers when it is weighed in water.

Fig. 284 c represents a bottle with a tube stopper having a contracted neck on which the normal quantity is marked. All the liquor above that mark is removed by a roll of filter paper, and the upper stopper prevents further evaporation.

Price of these Bottles, of the capacity of about half an ounce, but not adjusted : figs. *a* and *b*, 1s. each; fig. *c*, 2s.

285. *Estimation of the Specific Gravity of a Liquid by Weighing it in a Bottle.*—The specific gravity of a liquid can be determined by weighing a quantity of it in a bottle of the form and size represented by fig. 285. This bottle, when filled up to the mark *a a* on the neck, contains the hundredth part of an imperial gallon, or 700 grains of water at 62° Fahr. The bottle must have a brass counterpoise with a moveable cover, to admit of occasional corrections of the weight. Its convenient use also demands the following additions.

A pipette, fig. 285 *b*, of the capacity of 50 septems, for use in filling the bottle to *a a* without wetting the outside.

284 c.

A narrow pipette, for adjusting the liquor in the bottle to the normal mark *a a*, after regulating the temperature.

And a thermometer cased in glass, fig. 285 *c*, and made narrow enough to pass through the narrow neck *a a* of the bottle, fig. 285. This bottle may be weighed with the balance, No. 124, and the weights No. 127.

285. *Price of the articles named above* :—

	s.	d.		s.	d.
a. The Bottle ..	2	6	*d*. 3 Small Pipettes ..	0	6
b. The Counterpoise	2	0	*e*. Thermometer ..	4	0
c. Pipette, 50 septems	1	0	*f*. The Set	10	0

When an experiment is to be made, the bottle should be filled by means of the pipette a little above the mark *a a*; the thermometer should then be used to try the temperature. If found correct, namely at 60° Fahr. for alcohol, and at 62° Fahr. for all other liquids; then the small pipette should be used to adjust

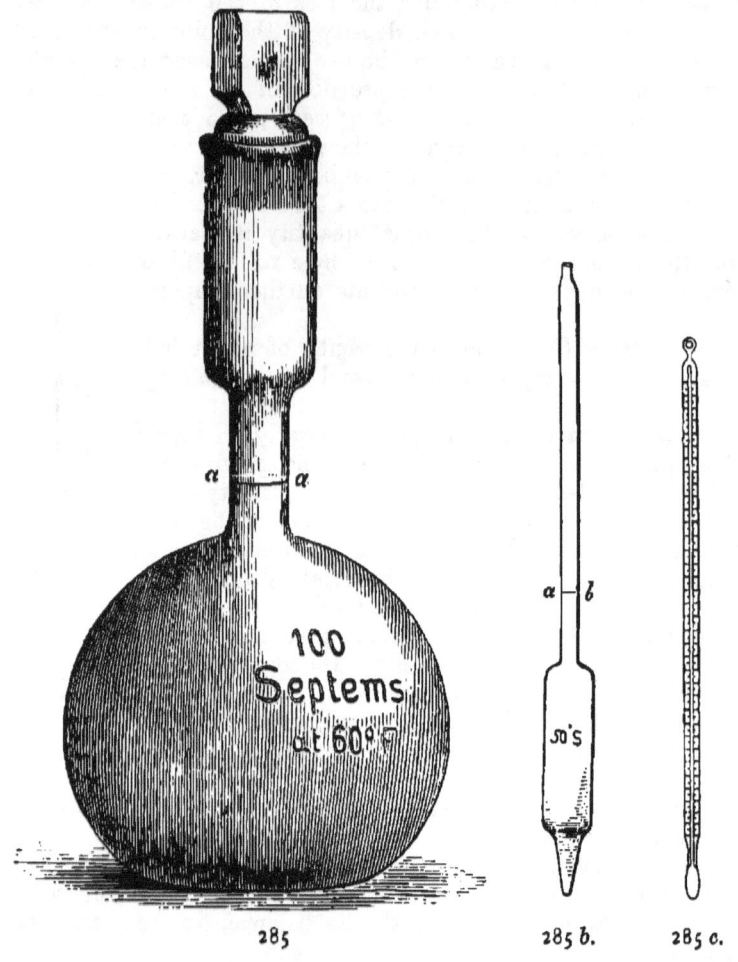

285.   285 b.   285 c.

the liquor accurately to the mark *a a*, and the vessel be closed by the stopper. It may then be weighed.

Deduct the weight of the counterpoise from the weight of the filled bottle, expressed in grains, and divide the residue by 7; the result is the specific gravity of the liquid submitted to trial, compared with water at 100.

SPECIFIC GRAVITY BOTTLES.

Various Forms of Specific Gravity Bottles for Liquids.

For various purposes of science, or manufacture, many alterations have been made in the form of specific gravity bottles, of a few of which it may be useful to give a brief description. Fig. 287 to 293 represent the most important varieties. We subjoin a list of Prices of different sizes of each kind.

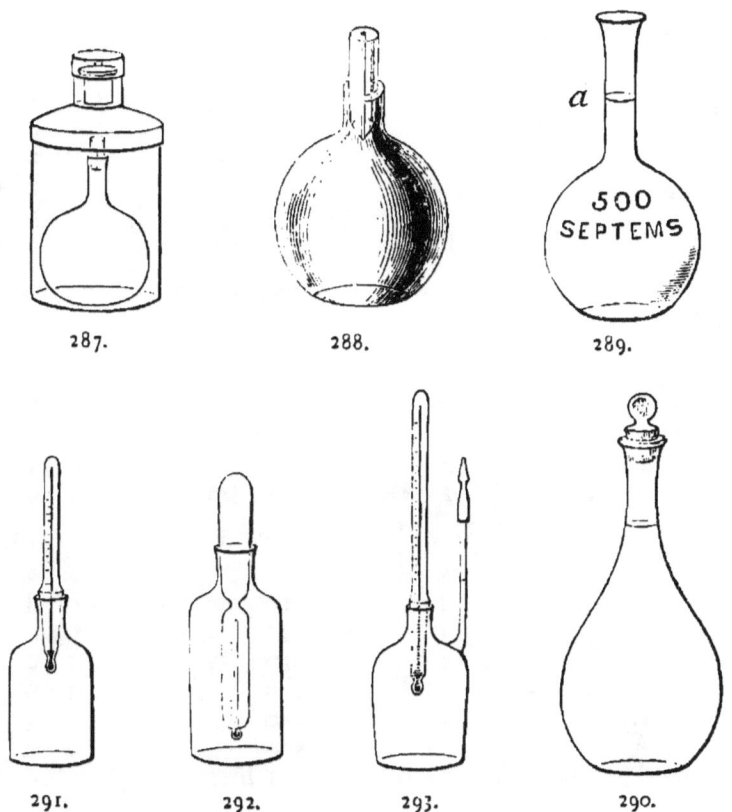

287.   288.   289.

291.   292.   293.   290.

285. *Regnault's Specific Gravity Bottle*, form of fig. 285, with stopper and mark on the narrow part of the neck :—

	s.	d.			s.	d.
250 grains	2	0	25 grammes ..	2	0
500 ,,	.. 2	3	50 ,,	2	6
700 ,,	.. 2	6	100 ,,	3	0
1000 ,,	.. 3	0			

287. *Slight Blown Glass Bottles*, with perforated stopper like fig. 288, in a japauned tin case with counterpoise, fig. 287:—

	s.	d.			s.	d.
250 grains	4	0	25 grammes	..	4	6
500 ,,	4	6	50 ,,	..	5	0
1000 ,,	5	0	100 ,,	..	5	6

288. *Slight Blown Glass Bottle*, with perforated stopper, without case or counterpoise, fig. 288; or
Solid Glass Bottle, fig. 288 A, with massive stopper, having a vertical groove on the side, without case or counterpoise:—

	s.	d.			s.	d.
100 grains ..	2	0	5 grammes		1	0
250 ,, ..	2	6	10 ,,		2	0
500 ,, ..	3	0	20 ,,		2	3
700 ,, ..	3	6	25 ,,		2	6
1000 ,, ..	4	0	50 ,,		3	0
1 cubic inch	2	0	100 ,,		3	6

288 A.

289. *Measuring Flasks*, without stoppers, with 1 mark on the neck:—

	s.	d.			s.	d.
50 septems	0	9	10 grammes	..	0	4
100 ,,	1	0	25 ,,	..	0	6
500 ,,	1	9	50 ,,	..	0	8
1000 ,,	2	0	100 ,,	..	0	10
1 pint	2	3	1 litre	..	2	0

290. *Measuring Flasks*, with stoppers, and 1 mark on the neck:—

	s.	d.			s.	d.
100 septems	1	0	10 grammes		0	8
250 ,,	1	6	25 ,,	..	0	9
500 ,,	2	0	50 ,,	..	1	0
1000 ,,	2	6	100 ,,	..	1	6
1 pint	3	0	250 ,,		1	9
1 cubic inch ..	1	0	500 ,,		2	0
1000 grains	1	0	1 litre		2	6

291. *Specific Gravity Bottles*, with thermometers ground in as stoppers.

292. *Specific Gravity Bottles*, different pattern, same price:—

293. *Specific Gravity Bottle*, 1 cubic inch, with finely divided thermometer, ground in as stopper, and an extra neck with 1 mark and ground cap. Price 10s. 6d.

294. *Solid Glass Specific Gravity Bottle*, with solid stopper, having a vertical groove on the side, form of figure 288 A; accurately justified with counterpoise in leather case:—

		s.	d.			s.	d.
100 grains		5	0	10 grammes	..	5	0
250 ,,	..	6	0	25 ,,	..	6	0
500 ,,	..	7	6	50 ,,	..	7	6
1000 ,,	..	10	0	100 ,,	..	10	0

295. *Practical Remarks on the Choice and Use of Specific Gravity Bottles.*—As there is always a chance of error in measuring a definite quantity of liquor to be weighed, and also in changing the temperature of the liquor while wiping the bottle after filling it, especially when a perforated stopper is used, and as such errors affect small quantities more seriously than large quantities, it is advisable in all cases to weigh the largest quantity of a liquor which the balance will carry, provided enough liquor is at command. For example:—8 fluid ounces of water (3500 grains) weighed in a balance that turns with $\frac{1}{4}$ grain (see No. 124), will give a more accurate result than 500 grains weighed in a much more accurate balance; because there may be in both cases an error of 1 grain in filling the bottle and finding the temperature, and that error falls less heavily on 3500 than on 500 grains.

Knowing the power of the balance that is to be used, it is easy, by means of measuring flasks and measuring pipettes, to measure off the greatest quantity of liquor that will suit the balance, and this can be done without giving rise to troublesome calculations. In proof of which I give the following tables, which embrace all the ordinary septem, decem, and gramme measures. By the word *septem* I mean the measure of 7 grains of water, and by *decem* the measure of 10 grains of water, at 62° Fahr.

The first Table gives, in the second column, the most common septem and decem measures: in the first column, their equivalents in grains of water at 62° Fahr.; and, in the third column, the number of times which the weight in grains must be taken to represent the specific gravity of the liquors that are tried in comparison with water fixed at 1000.

EXAMPLES.—1. 100 septems contains 700 grains of water. If this is divided by 7 and multiplied by 10, we have 1000 as the standard of specific gravities.

2. If a liquor $1\frac{1}{2}$ times the weight of water is weighed in the same bottle, it is clear that 100 septems will weigh 700 + 350 =

1050 grains. If this number is divided by 7 and multiplied by 10, we have 1500, which is the specific gravity required.

Grains of Water.	Septems or Decems.		Multipliers.		Grains of Water.	Septems or Decems.		Multipliers.
7	1	S.	$\frac{1000}{7}$		600	60	D.	$\frac{10}{6}$
10	1	D.	100		700	100	S.	$\frac{10}{7}$
35	5	S.	$\frac{200}{7}$		750	75	D.	$\frac{4}{3}$
50	5	D.	20		800	80	D.	$\frac{10}{8}$
70	10	S.	$\frac{100}{7}$		875	125	S.	$\frac{8}{7}$
100	10	D.	10		900	90	D.	$\frac{10}{9}$
125	12½	D.	8		1000	100	D.	0
140	20	S.	$\frac{50}{7}$		1250	125	D.	$\frac{8}{10}$
150	15	D.	$\frac{20}{3}$		1500	150	D.	$\frac{2}{3}$
175	25	S.	$\frac{40}{7}$		1750	250	S.	$\frac{4}{7}$
200	20	D.	5		2000	200	D.	$\frac{1}{2}$
250	25	D.	4		2500	250	D.	$\frac{4}{10}$
300	30	D.	$\frac{10}{3}$		3000	300	D.	$\frac{1}{3}$
350	50	S.	$\frac{20}{7}$		3500	500	S.	$\frac{2}{7}$
400	40	D.	$\frac{10}{4}$		5000	500	D.	$\frac{2}{10}$
500	50	D.	2		7000	1000	S.	$\frac{1}{7}$

Grammes.	Multiplier.	Grammes.	Multiplier.
5	200	100	10
10	100	200	5
20	50	250	4
25	40	500	2
50	20		

3. A cheap laboratory balance, such as No. 124, which will carry 8 ounces in each pan and turn with $\frac{1}{20}$ grain, will give good general results with a bottle of the form of fig. 290, and of the capacity of 250 septems = 4 fluid ounces; since even this quantity of concentrated oil of vitriol will be within the range of the balance, for if the specific gravity of the acid is 1·846, 250 septems of it will weigh 3230·5 grains. According to the Table the equivalent of water is 1750 grains, and if that is divided by

7, and multiplied by 4, it gives 1000. So also, 3230·5, if divided by 7 and multiplied by 4, gives 1845 as the required specific gravity.

The *second Table* gives the factors to be used in reckoning the products of weighings of gramme (or centimetre cube) measures expressed in gramme weights.

296-298. *Estimation of the Specific Gravity of a Liquid by Weighing in a Beaker a quantity first measured by a Pipette.*

296. The specific gravity of any acid, alkaline, or saline solution, can be readily tested by the following process. By means of a measuring pipette, fig. 296, transfer a small carefully measured quantity of the solution that is to be tried, into a beaker, fig. 297, or a wide-mouthed light flask, fig. 298, and *weigh* it. To save trouble in calculation the quantity thus measured should be one of the quantities named in the above Table; say, for example, 100 grains; then, the weight of the liquid, multiplied by 10, gives the specific gravity.

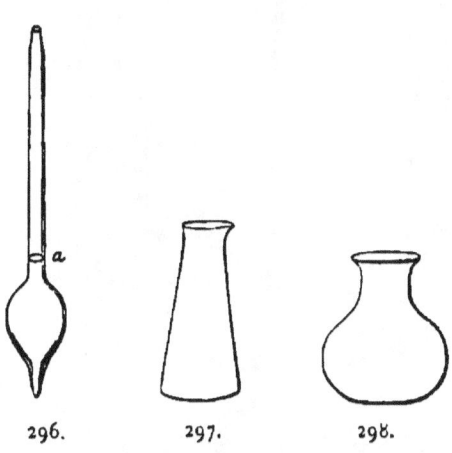

296.   297.   298.

	s.	d.
Price of pipette, fig. 296 = 1000 grains	1	3
„ jar, 297 = 5 ounces	0	8
„ flask, 298 = 5 ounces	0	5

299. *Mohr's Hydrostatic Balance,* for taking the specific gravity of liquids by an easy method, which gives the specific gravity without calculation, and requires but a small quantity of liquid, and no other apparatus than that represented in fig. 299.

299. *Price, in a mahogany glass case, including a portable mahogany box,* 4*l.* 14*s.* 6*d.*

300. *Price, in a mahogany box, without glass case,* 2*l.* 15*s.*

Description of Mohr's Specific Gravity Balance.—This apparatus consists of the articles represented in fig. 299; namely, a 10-inch beam, of which 1 branch is divided into 10 parts; a glass plummet, which contains a thermometer, and is attached to a platinum wire by which it can be suspended from the beam; a glass cylinder,

HYDROSTATICS.

and a mahogany tray for it; a small brass pan to counterpoise the plummet; a pair of pans, marked $e\ e$ in the figure, for the ordinary weighing of solid bodies; a set of riders, marked $a\ b\ c\ d$, of which d is equal to the weight of the water displaced by the plummet, while $c = \frac{1}{10}$ of d, $b = \frac{1}{10}$ of c, and $a = \frac{1}{10}$ of b. The apparatus can be dismounted and packed in the mahogany box, and then removed from the glass case for travelling.

The scale of the thermometer included in the glass plummet is usually that of Réaumur, for which reason we give the following equivalents of the degrees most likely to be observed:—

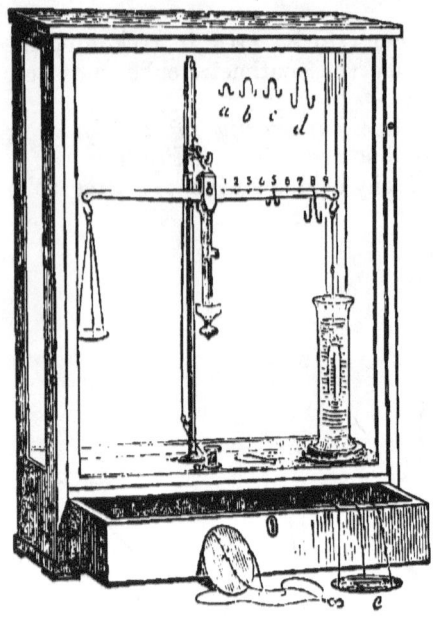

299.

R.	F.	R.	F.	R.	F.	R.	F.
20°	77·°	16°	68·°	12°	59·°	8°	50°
19	74·75	15	67·75	11	56·75	7	47·75
18	72·5	14	63·5	10	54·5	6	45·5
17	70·25	13	61·25	9	52·25	5	43·25

Process.—1. When the small pan is attached to one end of the beam and the plummet to the other end, the beam rests in equilibrium. 2. If the plummet is plunged into distilled water, that end of the beam rises. 3. If in that case one of the largest riders, d, is put on the hook at that same end of the beam, the equilibrium is again restored. This result shows that the rider d is equal in weight to the displacement of the plummet.

4. If a liquid lighter than water is submitted to trial, the large rider d must be placed somewhere on the divided branch of the beam, where it will produce equilibrium. But when, as commonly happens, this occurs at some point between two notches on the beam, it is best to rest the rider on the notch of lowest value of the two between which it rests, and then to apply the next sized rider, c, to determine the difference. Thus, in fig. 299, we see the

SPECIFIC GRAVITY BALANCES. 71

large rider at 8, and the small rider at 5. In this case, the former represents 8 in the first place of decimals, and the latter represents 5 in the second place of decimals. Thus, 0·85. If the rider c does not effect an equilibrium, when fixed in a notch, but falls between 4 and 5, then it must be placed in 4, and the third sized rider, b, must be employed to find the exact point of equilibrium. This third size of rider stands for the third place of decimals in the expression of the specific gravity: and finally, the smallest rider, a, expresses the fourth place of decimals, thus:—

a is = ·0001 to ·0009
b is = ·001 to ·009
c is = ·01 to ·09
d is = ·1 to ·9

} According as they stand in the notches 1 to 9 on the beam when at rest.

5. If a liquid heavier than water is tried, the process goes on exactly as above described, with the addition that one of the riders d is in every case hung at 10, the extreme end of the beam, to serve as the equivalent of the weight of water, while the riders which show the difference between water and the given heavy liquid cross the beam at the proper points. A few examples will illustrate this principle. In the following plans the figures represent the notches on the beams, and the letters the respective sizes of riders:—

```
1   2   3   4   5   6   7   8   9   10
    b       c               d       d   = 1·842 Sulphuric Acid.
    dc                              d   = 1·33  Caustic Potash.
                    c               d   = 0·96  Ammonia.
                c       d               = 0·75  Ether.
    d   b                       c   d   = 1·495 Nitric Acid.
```

Power of the Balance to weigh Solids.—It will carry 1000 grains, and show $\frac{1}{10}$ grain. With 500 grains, it shows $\frac{1}{20}$ grain; and with 300 grains, it shows $\frac{1}{30}$ to $\frac{1}{50}$ grain. The larger pair of pans, marked e in fig. 299, are used for weighing solids.

301. *Mohr's method of estimating the Specific Gravity of a Solid by measuring the Water it displaces.*—The apparatus consists of a glass beaker, ground smooth on the edges; a flat slip of wood or brass, having, on the under side, a conical piece of brass ending in a fine point; and a pipette. See fig. 301. *Price* 6s.

When the experiment is to be made, the beaker must be fixed by a little wax upon a steady horizontal table. The conical brass spike must be rubbed with chloride of platinum, which makes it black, and afterwards be slightly greased with tallow. If the solid body that is to be examined is weighed in grammes, the pipette must be graduated into centimetre cubes. If the solid is weighed in grains, the pipette must be graduated into grains or septems convertible into grains.

Process.—Pour some water into the beaker, put on the brass

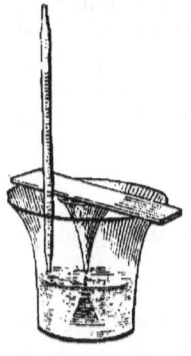

301.

pointer, and add water to the beaker, carefully and slowly, by means of the pipette, until the surface comes into contact with the brass spike. This can be done with wonderful precision; for when the light of the sky plays upon the surface of the water, it forms a brilliant white ground upon which the reflection of the black spike is very distinct, and the meeting of the water and index is instantly recognised. The index is then removed, the substance whose specific gravity is to be estimated, and which has already been weighed in air, is put into the water; and the pipette, which must be previously wetted inside with other water, is to be used to take out of the beaker as much water as appears to be a little more than the volume of the solid body under trial. The index is then to be replaced on the beaker, and, if the proper quantity of water has been taken up by the pipette, the index will not touch the water in the beaker. Then water is to be slowly dropped from the pipette into the beaker, until the surface rises to touch the black spike. *The water that then remains in the pipette is necessarily the volume of the solid body.* The value is read off in grammes or grains, according to the graduation of the pipette; and, on dividing the weight of the solid by that of the water, we have the required specific gravity.

302. *Nicholson's Hydrometer, for taking the Specific Gravity of Minerals and other Solids.*—Fig. 302.

A. *Price, in japanned tin,* 5s. B. *Price, in brass,* 6s. 6d.

This instrument is formed as shown by fig. 302. It is usually about 9 inches long and $1\frac{1}{2}$ inch in diameter. At the top is a small basin for holding weights or the solid substance that is to be tried. There is a small pan at the bottom, which is loaded with lead to make the instrument swim in a vertical position. In the middle of the neck a mark is painted or cut (not shown in the figure). A normal weight, say 200 grains, when put into the upper basin, sinks the instrument in pure water at 62° Fahr. down to the mark on the neck. That is the normal condition of the instrument.

When an experiment is to be made, the instrument is to be placed in water at the temperature of 62° Fahr., and the solid to be tried is to be put into the upper basin, with as many weights as will make up the 200 grains which constitute the normal power of the instrument. The weight being completed, the instrument sinks till the surface of the water cuts the mark on the neck. Deducting from 200 the weights added to the solid, the residue is the weight in grains of the solid when weighed in air. Raise the hydrometer in the water, remove the solid from the upper basin to the pan at the bottom, and sink the instrument again in the water, when it will be found to have lost weight, which necessarily is the weight of the volume of water displaced by the solid (see § 281); then add to the weight that remains on the basin as much more as will restore the equilibrium. The weight so added is equal to that of the displaced water. Dividing the weight of the solid in air by the loss of weight in water, the quotient is the specific gravity.

Example—In an experiment with a piece of sulphur the weight in air was found to be 64·2 grains, and the loss of weight in water was 31 grains. Hence 64·2, divided by 31 = 2·07.

The Hydrometer.

Estimation of the Specific Gravity of Liquids by means of the Hydrometer.

310. The hydrometer is an instrument employed to determine the specific gravity of liquids. It is commonly made of glass, especially when it is to be used with corrosive chemical liquids; but, for a few special purposes, it is made of brass or silver. The usual forms of the glass instruments are represented by figs. 310 a, b, c, d, e, f, g. Every instrument in the course of its manufacture is weighted at the bottom with mercury or lead shot, to make it sink to a certain depth in water and to make it swim in a vertical position; and the neck of every instrument contains a scale of figures to indicate the depth at which it swims in any particular liquid.

311–314. Figs. 311, 312, 313, 314, represent some of the forms of glass vessels that are used to hold the liquids that are to be tried by hydrometers.

It is evident that an instrument of this description, put into a liquid, will sink far down in a light liquid, and only a little way down in a dense liquid; since in every instance the weight of the quantity of liquid displaced is equal to the weight of the instrument. Accordingly, the depth to which the instrument sinks in the liquid under trial is indicated by a certain number on the scale inserted in the spindle or neck of the instrument, and from that number, or *degree*, as it is commonly called, a judgment is formed of the value of the substance that is held in solution in the liquid.

HYDROSTATICS.

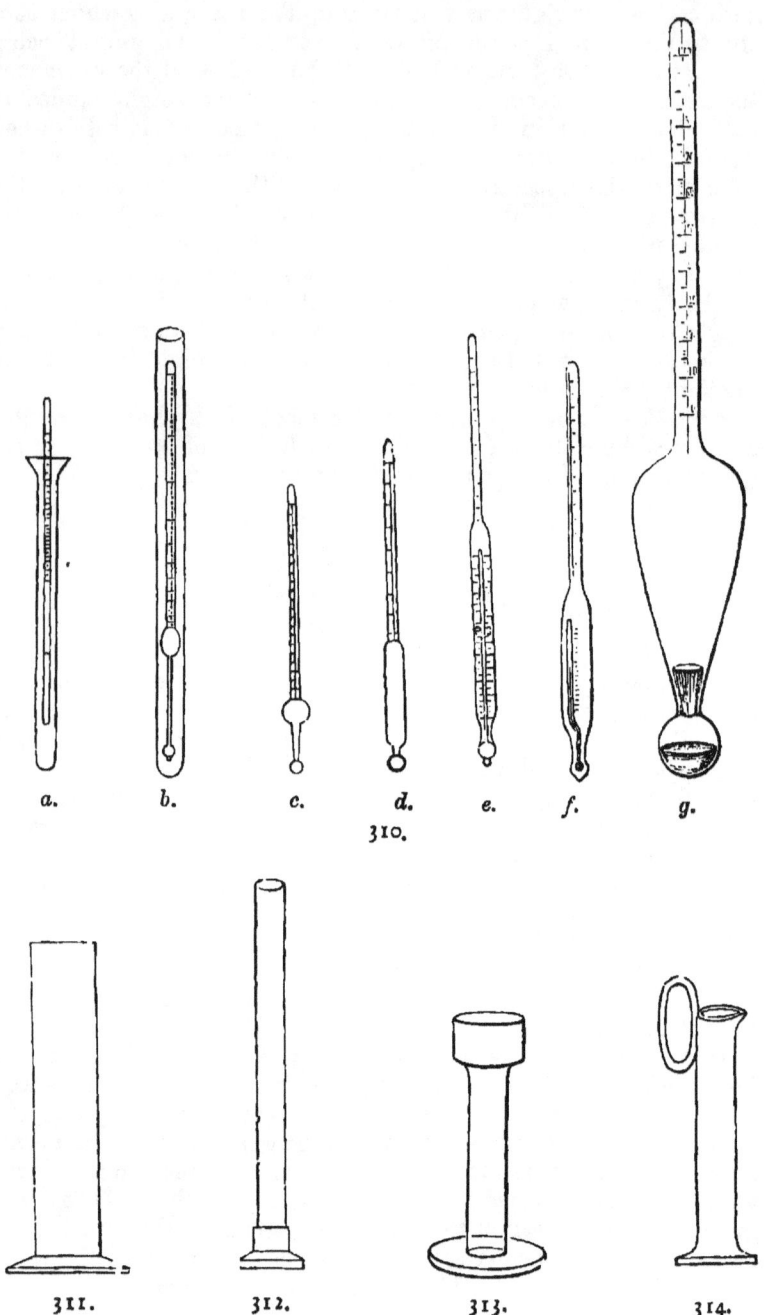

a.   b.   c.   d.   e.   f.   g.

310.

311.   312.   313.   314.

THE HYDROMETER. 75

315. It may be useful to give in detail an example by way of showing the difference in usefulness between the methods, already detailed, of estimating the densities of liquids, and this method of using the hydrometer. I will take the case of *spirits*, or mixture of alcohol and water.

If you fill a bottle of the exact contents of No. 285 with pure water and weigh it at 60° Fahr., you will find the weight of the water to be 700·00 grains. If you divide this number by 7 and remove the decimal point from 2 places to 4, the product will be 1·0000, which number may be taken to express the standard specific gravity of water when it is to be compared with the specific gravity of alcohol.

If you perform the same operation with absolute alcohol, the weight will be found to be 556·23 grains, and the specific gravity to be ·79461.

Between these two liquors a hundred mixtures can exist, differing from one another in a regulated degree; beginning, for example, with a mixture containing 1 per cent. of alcohol and 99 per cent. of water, and ending with a mixture containing 99 per cent. of alcohol and 1 per cent. of water. Each of these hundred mixtures could be weighed in the bottle No. 285, and its specific gravity could be determined by the process just explained. I subjoin a few examples, extracted from a complete table of such trials, given in *Griffin's* 'Chemical Testing of Wines and Spirits,' London, 1872, page 33:—

Percentage of absolute Alcohol by volume.	Weight in grains of 100 Septems of the spirit at 60° F.	Specific Gravity at 60° F. Water = 1·00000 at 60° F.
0	[Water = 700·00]	[Water = 1·00000]
1	698·95	·99850
2	697·90	·99700
5	694·95	·99279
10	690·61	·98659
50	654·04	·93434
57·06*	643·89	·91984
99	559·66	·79952
100	556·23	·79461

* Proof spirit, according to Sikes.

316. This method of ascertaining the strength and value of a mixture of alcohol and water is very trustworthy and precise; but it requires fine instruments—a good balance, and accurate weights

and measures: it demands much care and occupies much time. The same may be said of the solutions of other substances than alcohol. These are serious considerations; and in business, whether the business be manufacturing or commercial, men demand quick and cheap modes of working out solutions of scientific problems, even at the sacrifice of a little accuracy. For this reason they take the strength of spirits, and of many other substances in solution, by means of the hydrometer. The process is very simple. The spirit to be tried is put into a glass jar (see figs. 311 to 314), which is filled nearly to the top. The hydrometer, or alcoholometer as it is called when used for spirits and provided with a suitable scale, is held by the upper end and gently let down into the spirit. After a little oscillation it settles in the liquid, and the degree on the scale where it cuts the surface of the liquid shows its quality or strength. If the instrument is made like fig. 310 g, with a scale of 100 degrees corresponding with the 100 mixtures of alcohol and water, already explained, then the degree indicated by the action of the instrument shows the percentage of alcohol contained in the mixture under trial.

317. *Peculiarities of the Forms of Hydrometers, represented by fig.* 310.—a shows an instrument of the simplest form, without a bulb. This form can be used with a small quantity of liquid, but it is not very accurate, because the spindle is necessarily rather thick and the instrument suffers from the capillary attraction of the solution. The funnel formed at the top of the solution tube serves to prevent the adhesion of the spindle to the side of the tube. The solution must fill the tube up into the funnel. b. With a small bulb the instrument can swim without adherence to the tube, yet with a small bulb and a large spindle the instrument remains comparatively inactive. c. A spindle with a bulb. More accurate than the preceding forms. The larger the bulb, and the longer and thinner the spindle, the more accurate is the instrument. d. Needs the same remarks as made on c. Forms e and f differ from d only in having thermometers enclosed in their tubes. Form g swims rather more sensitively than c and d.

318. *Forms of Solution Jars.*—Fig. 311 is used for pretty large instruments where there is plenty of liquor. 312 is used when smaller quantities of liquor have to be tried. It is also the form chosen to put into sets of portable hydrometers. It has a heavy metal foot. 313, filled halfway up the upper cup, serves for all instruments, and the cup is useful to prevent the adhesion of tubes to the side of the jar. 314 represents a stoneware jar, such as is used by workmen in chemical works to dip from the vats a sample of solution, to be tried by the hydrometer, so as to ascertain how near it is to the quality required for the next manufacturing operation.

319. *Varieties of Hydrometer Scales.*—Hydrometers are of continued use in commerce and in chemical manufactures, and they are fitted with a variety of scales according to the purposes for which they are used. It will suffice to name a few of them.

320. For *general purposes*, the scales simply indicate *specific gravity* against water taken as a standard, and marked 100 or 1000. The scale is sometimes put upon one spindle, having then the form of 310 *a* or *b*; sometimes it is divided upon as many as seven spindles running from 700° up to 2000°, by degrees equal to 0.001. Fig. 320 represents a set of four spindles arranged in a box, with solution tube and thermometer.

320.

Twaddell's Hydrometers.—Sometimes the long numbers of the degrees on scales of this kind are found troublesome in practice, especially when workmen are to use the instruments. To obviate this defect Twaddell, making an instrument in six spindles, allowed the degrees to be larger, so that from sp. gr. 1000 to sp. gr. 1950, he had 190°, each degree being equal to ·005. This scale is now much used in Glasgow and Manchester. To convert Twaddell's degrees into specific gravity, multiply Twaddell by ·005 and prefix 1. Twaddell's hydrometer is not graduated for light liquids.

Baumé's Hydrometer.—Baumé made two scales, one for heavy liquids and one for light liquids. Many interpretations have been given of the value of his degrees. The following are as nearly correct as probably can be given. Let B = Baumé's degree, and 100 = water, then:

For *heavy liquids*, specific gravity $= \dfrac{144}{144 - B}$.

For *light liquids*, specific gravity $= \dfrac{128}{118 + B}$.

321. *Special Scales* in general use, are such as indicate the *percentage* of some given ingredient—alcohol, sugar, potash, soda, ammonia, sulphuric, hydrochloric, and nitric acids, &c. These scales are necessarily founded on an accurate chemical knowledge of the compositions of the solutions of these respective substances. On the Continent, Baumé's scales are used for many of these compounds; but in England and Scotland Twaddell has displaced Baumé.

Among special scales for hydrometers may be cited those made for hydrometers that are to be used in very hot climates, such as the West Indies. This is especially the case with alcoholometers and saccharometers. The instruments made for use in Europe are graduated at 60° Fahr. for spirits, and at 62° Fahr. for sugar; but both require to be used in the West Indies at about 84° Fahr. To overcome that difficulty the instruments are specially graduated when made in London, so as to act correctly when used at 84° Fahr.

322. *Rules to be observed in using Hydrometers.*—The instrument must be clean. It should be washed after every experiment, and be wiped dry and put into its case. When it is to be used, it must not be fingered, even with clean fingers, and still less with greasy fingers. It should be held by the extreme upper end, and be let down slowly into the liquor to be tried. It must swim vertically in the liquor, otherwise it is useless. There must be no air-bubbles either on the hydrometer or on the sides of the trial jar. The hydrometer must be wet with the solution from the bottom of the instrument up to the point where it cuts the surface of the solution. It must not be wetted higher than that point. Great care should be taken always to watch in the same manner the point where the spindle meets the surface of the water. The temperature of the liquor should be taken before the hydrometer is sunk into it, and before the thermometer is used the liquor should be well stirred with a clean glass rod, in order that different horizontal sections of it may not retain different temperatures.

323. Price-list of a few Hydrometers.

324. Hydrometer, one glass spindle, with solution tube. It indicates approximately all specific gravities from alcohol to oil of vitriol, or from 0·70 to 2·00, water being marked 1·00. *Price 6s.*

325. Ditto, same range, on two spindles. *Price 6s.*

326. Ditto, 700 to 2000, on three spindles. *Price 10s.*

327. Hydrometers, a set of four spindles, with thermometer, and a solution tube arranged in a mahogany box, lined with leather, like figure 320. *Price 30s.*

328. Hydrometers, a set in a box, similar to No. 327, but extending to seven long spindles. A very delicate set. *Price 42s.*

329. Twaddell's hydrometers, form of fig. 310 g; set of six, in a mahogany box. *Price 15s.*

330. Twaddell's hydrometers, single spindles Nos. 1 and 2, each 1s. 6d. Nos. 3 and 4, each 1s. 9d. Nos. 5 and 6, each 2s.

331. Alcoholometer, according to Tralles, 100°, with thermometer. *Price 3s. 6d.*

332. Ditto, 100°, without ditto. *Price 2s.*

333. Alcoholometer, according to Gay-Lussac, 100°, with thermometer. *Price 3s. 6d.*

334. Ditto, 100°, without ditto. *Price 2s.*

FLOTATION. 79

335. Sikes's alcoholometer, scale 60° O. P. to 40° U. P.
a. Scale on paper, 2*s*. *b*. Scale on ivory, 2*s*. 6*d*.

336. Hermbstadt's saccharometer, one glass spindle, with two scales, namely, specific gravity from 1000 to 1321, and percentage of sugar from 0 to 67. *Price* 5*s*.

337. Lactometer for testing milk, showing additions of $\frac{1}{4}$, $\frac{1}{2}$, and $\frac{3}{4}$ water. *Price* 1*s*. 6*d*.

In *Griffin's* 'Chemical Handicraft' is given an extended list of varieties of hydrometers, alcoholometers, saccharometers, urinometers, and other instruments of this description.

FLOTATION.

345. Apparatus for estimating the weight of floating bodies by the measurement of the water they displace, consisting of
a. Circular glass basin, 8 inches diameter. *Price* 2*s*. 6*d*.
b. Brass gauge, to measure the height of the water. *Price* 2*s*.
c. Pipette. *Price* 9*d*.
d. Glass measure, graduated into grains. *Price* 3*s*.
e. Section of a small ship. *Price* 2*s*. 6*d*.
The set, 10*s*. 6*d*.

Process.—This experiment is founded on that of Dr. Mohr, given at § 301, the details of which may now be consulted.

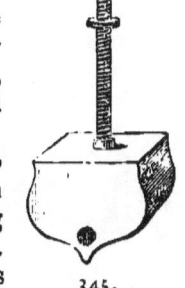

345.

Suppose in the present case the problem to be solved is to determine the weight in grains or ounces of a model of a transverse section of a ship, as represented by fig. 345, and which we will assume to be about 4 ounces.

The basin, *a*, is to have so much water put into it as will just meet the point of the gauge, *b*, when that is put across the basin. The method of doing this with accuracy has been fully explained at § 301. The gauge is then to be removed and the ship is to be put into the water, in which there will be a considerable rise. A quantity of water is to be transferred by means of the pipette into the graduated cylindrical jar, a quantity rather more than is equal to the weight of the model. The test of this is, that when the gauge is returned to its *exact position* on the basin (which must be carefully attended to), the metal point or index does not touch the surface of the water. The adjustment must then be made by transferring water in drops slowly by the pipette from the graduated jar into the basin, until the water touches the point of the index. The residue of water in the pipette is then to be added to that in the graduated jar, and the total quantity is to be read off on the graduated scale. If the quantity is expressed in grains or ounces, *it is the weight of the model*, in agreement with the law of Archimedes, as explained in § 271 c, viz., that "when a solid swims on water, its weight is equal to that of the water which is displaced by its submerged portion."

During this operation the position of the basin must be unaltered, the index must always rest in the same position, the pipette must be wetted inside with other water before commencing the operation, and there must be no loss of water in the transfers.

346. The conditions of equilibrium of a floating body may be conveniently illustrated by experiments made with the little model ship, fig. 345. and the water basin 345 a. The ship has a mast in the centre of the deck, and upon this a brass screw works from end to end. The keel of the ship is so loaded by a bar of lead, that when the brass weight is at the bottom of the mast the model, when floating in the water basin, will right itself after being displaced laterally; but when the brass weight is screwed up to the top of the mast, the model will readily upset. The changes in the state of equilibrium, and the passage from stability to instability, may be examined by gradually shifting the position of the brass screw.

HYDRODYNAMICS.

351. The science of *Hydrodynamics* treats of liquids in a state of movement. A special department of this science is devoted to the art of conducting and raising water, and is called *Hydraulics*.

CAPILLARITY.

352. The different conditions of equilibrium which have been illustrated in the preceding chapters suffer remarkable exceptions when the liquids are contained in very narrow vessels, or when we consider the action of the liquids in contact with the sides of the containing vessels. We can only refer to two or three particulars.

Water, when poured into a glass vessel and shaken a little, wets the sides of the glass above the horizontal level of the mass of water, and the surface of the water assumes a concave figure.

When a glass tube is dipped into water, the water rises round the outer sides of the tube to a higher level than the horizontal surface of the water.

Inside the tube the water rises to a level more or less above the level of the water outside the tube, and assumes a concave surface.

The height to which water thus rises in a tube increases as the diameter of the tube diminishes. In short, the elevation is inversely proportional to the diameter of the tube.

Different liquids have different ascensional powers. Experiments made at $65°$ Fahr. with a tube of $\frac{1}{25}$ inch diameter, give the following results:—Water rose about $1\frac{1}{2}$ inch, alcohol $\frac{1}{2}$ inch, nitric acid 1 inch, and essence of lavender $\frac{1}{8}$ inch.

Mercury, when poured into a glass vessel, recedes from the sides of the glass, leaving a space between the glass and the horizontal surface of the mercury.

When a glass tube is dipped into mercury, the metal is repelled by the outer surface of the glass.

Inside the tube, the mercury is kept down below the external horizontal level; it is repelled from immediate contact with the glass, and acquires a convex surface.

The student is recommended to try these experiments with a little clean mercury and a variety of narrow glass tubes.

353. Set of six *capillary Glass Tubes*, fig. 353, with japanned tin frame and trough, for the exhibition of capillary attraction. Price 2s. 6d.

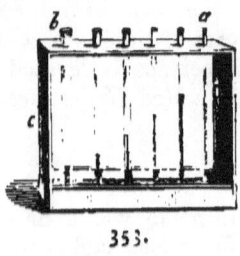
353.

354.

354. Pair of *flatted Glass Plates*, 4 inches by 3 inches, to show capillary attraction between surfaces, mounted in a japanned tin frame. Fig. 354. Price 2s. 6d.

355. Two *Wooden Balls*, 1 inch in diameter, to show capillary attraction when they float in water. *The pair, price* 4d.

All these capillary instruments, when put into water, must be particularly clean, and the tubes and plates should be wetted with distilled water, some of which also should be put into the little troughs.

356. *Capsule of Iron-wire Gauze*, which, when slightly oiled, contains water, and floats on water, showing the repulsion between water and greased surfaces. Price 1s. 6d.

357. *Filtering Tube*, acting by capillarity, 6 inches long, $\frac{1}{8}$ inch bore, smooth at both ends. Fig. 357. *Three for* 6d.

It often happens in chemical researches that when a turbid mixture has been well shaken, especially after boiling, the precipitate it contains slowly settles down and leaves a very shallow stratum of clear liquor upon its surface. By using a very narrow glass tube, cut off smooth and square at the ends, it is possible to remove from the surface of such a mixture a sufficient quantity of clean liquor for testing, without waiting till the mass of precipitate subsides. The method is exhibited by fig. 357. The tube is made barely to touch the surface of the liquor: it is not dipped into it; no suction is applied, because the effect of capillary attraction is alone sufficient to raise the required quantity of clear liquor. When the mixture allows this abstraction of clear liquor to be effected, it is, of course, much preferable to filtration.

357.

358. *Filtering Paper*, a stout roll put into a tall narrow jar, with some water coloured blue by indigo dye, No. 204 *b*, affords a good example of the ascent of liquids by capillary attraction.

358. *a.* Filtering paper, per quire. *Price* 1s. 6d.
  b. Jar, on foot, 10 by 2 inches. *Price* 1s.

PRESSURE OF LIQUIDS PROPORTIONATE TO DEPTH.

359. Particles of fluid, on escaping from an orifice, possess the same velocity as if they had fallen freely *in vacuo* from a height equal to that of the fluid surface above the centre of the orifice.—*Torricelli.*

360. *Spouting Jars for illustrating Torricelli's Theorem.*—Japanned tin-plate, form of fig. 360, size 20 inches high, 6 inches diameter, with necks of glass tube, and stoppers.

A. With two necks. *Price* 6s.
B. With three necks (fig. 360). *Price* 7s. 6d.

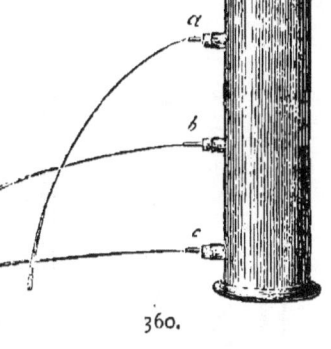

360.

If the vessel (fig. 360) is filled with water up to the mark D, and the apertures, *a, b, c,* are afterwards opened, the water will escape from them with very different velocities. At *a* the water will possess the same velocity as if its particles had fallen *in vacuo* from D to *a*; at *b* and *c* the escaping currents will possess the same velocities as if the liquid composing them had fallen from D to *b*, and from D to *c*.

Observe that the metal necks of the apparatus are closed by corks, that the corks are perforated and pierced by small glass tubes, and that these tubes are closed by small stoppers of caoutchouc or cork till the time comes for shewing the experiments.

THE PIPETTE.

361. *The Pipette* has been long used in chemical laboratories for transferring small quantities of different liquids from vessel to vessel. During the last thirty years it has derived increased importance in consequence of its great utility in the art of Volumetric Analysis. Figs. 361 to 367 represent some of the principal forms of the pipette. They are all open at both ends. The opening at the lower end is about $\frac{1}{20}$ inch in diameter. The opening at the upper end should be about $\frac{1}{6}$ inch in diameter, the glass there should be about $\frac{1}{10}$ inch in substance, and the end should be ground flat. The pipette should be held near the upper end by the thumb and middle finger of the right hand, while the fore-finger should be

84 HYDRODYNAMICS.

slightly wetted and kept ready to close or open the upper orifice of the instrument. When the pipette is to be filled, the lower end of

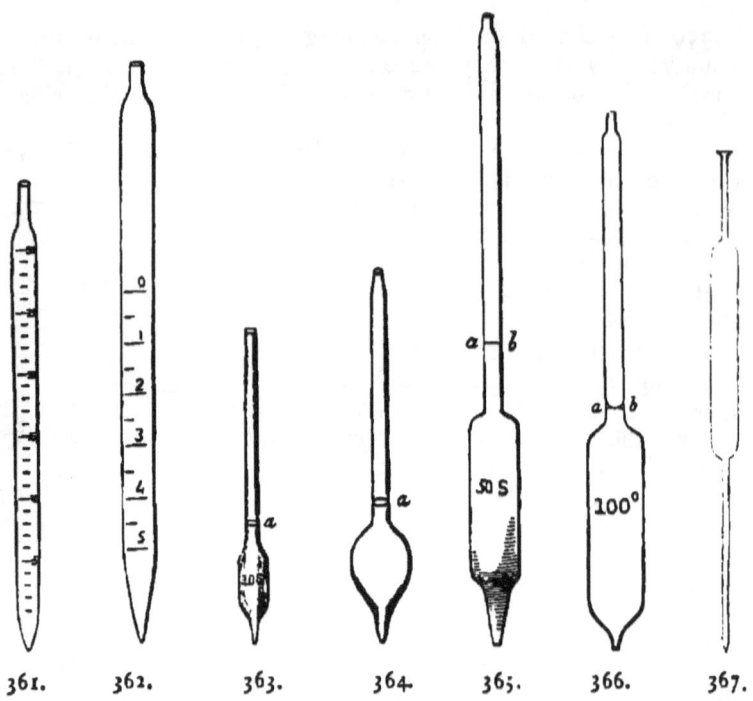

361. 362. 363. 364. 365. 366. 367.

it is to be put into the liquid and the mouth applied to the upper end to suck up the liquid. As that is sometimes poisonous, you must take care not to suck it into your mouth. Some authors recommend the pipette to be formed like fig. 368, with a bulb near the upper end, assuming that the liquid will, in case of need, rest in that cavity; but that is really an insecure safeguard, for if the lower end of the pipette rises suddenly above the surface of the liquid, then the air rushes through the liquid in the pipette with such velocity as to force it into the operator's mouth. A safer plan is to mount any short pipette with a caoutchouc tube, as represented by fig. 369. The glass tube is held by the left hand at a, the liquid is sucked up till it is visible

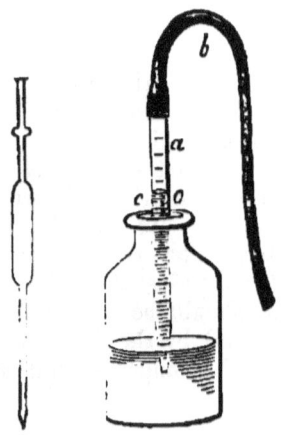

THE PIPETTE.

is pinched by the finger and thumb of the right hand at *b*, and the liquid secured in the pipette. Another plan, which a steady operator commonly finds sufficient, is to provide pipettes that are sufficiently long above the highest graduation, to enable him to see distinctly the height to which he can safely suck up the liquid. Keeping his eye upon that mark he fills his pipette, and then quickly closes the upper end with his finger.

In all cases the pipette is first filled above the mark by which the quantity for delivery is regulated; then holding the instrument in a vertical position the wet fore-finger is to be slightly relaxed to let air enter and cause some of the liquid to fall, drop by drop, till the surface in the pipette meets the required graduated line. The liquid in the pipette is then ready for delivery.

The pipettes, Nos. 363, 364, 365, 366, are graduated with one mark, *a*, or *a b*, and when filled up to that mark, they will deliver at once the marked number of measures of any standard description, say 10, 20, 50, or 100 septems, decems, centimetre cubes, &c. On the contrary, the measuring pipettes (figs. 361, 362) take up a quantity which, after adjustment, is to be delivered in partial quantities of 1, 2, 5, 10, measures, &c., as the analysis may demand. For the accurate delivery of very small quantities, the pipette, No. 362, is the best to use, because, when it has been carefully filled and justified at $0°$, it is possible to deliver one or two or five measures accurately, which cannot be done with the pipette, No. 361, in consequence of the uncertain delivery of the lowest degree. Pipettes of the form of figs. 367, 368, are commonly used for taking liquids out of long bottles that have narrow necks. Finally, the simplest form of pipettes is a glass tube of uniform bore, and with both ends cut off square. A use of such a pipette is shown in § 357. But a tube with uniform bore, and no confined jet at the bottom, cannot be safely used for the delivery of very small measured quantities of liquid, nor will it safely carry large quantities.

When mercury is to be lifted by a pipette, the bore of the pipette should not exceed $\frac{1}{3}$ inch; the substance of the glass should be nearly as much, and the upper end of the tube should be well ground, that the finger may have a firmer pressure.

PRICES OF PIPETTES.

Pipettes not graduated.

361. Plain pipette, form of fig. 361, 6-inch.   *Price* 2*d*.
364. Bulb pipette, form of fig. 364, 2 ounces.   *Price* 4*d*.
     Bulb pipette, form of fig. 364, 4 ounces.   *Price* 8*d*.
     Cylinder pipettes same price as bulb pipettes.

365. *Bulb Pipettes, graduated to deliver one quantity.*
Form of figs. 363 to 366.

	s.	d.		s.	d.
1 ounce	1	0	100 septems }	1	3
1 cubic inch ..	1	0	$= \tfrac{1}{100}$ gallon }		
100 grains ..	0	9	25 grammes ..	0	10
500 grains ..	1	0	50 grammes ..	1	0
1000 grains ..	1	3	100 grammes ..	1	3
50 septems ..	1	0			

361. *Scale of Pipettes*, namely, pipettes that are graduated from end to end, like figs. 361 and 362, to afford the means of measuring *any given quantity* within the capacity of the instrument:—

	s.	d.		s.	d.
1 ounce, in 8 drachms	2	0	50 septems, in 50 spaces	2	0
1 cubic inch, in tenths ..	2	0	100 septems, in 100 spaces	3	6
5 cubic inches, in fifths	2	6	25 grammes, in 50 spaces	2	0
100 grains, in 100 spaces	2	0	50 grammes, in 100 spaces	3	3
500 grains, in 100 spaces	3	0	100 grammes, in 100 spaces	4	6
1000 grains, in 100 spaces	3	6			

370. The *Burette*, or *Pouret*, is a modification of the pipette, having for object to reserve all the powers of the pipette as regards the transvasing and measurement of liquids, and at the same time to set free the operator's hands and permit him to apply his tests leisurely.

Fig. 370 represents Gay-Lussac's form of burette, adjusted to an apparatus for showing the exact height of the liquor in the instrument. Fig. 371 is Binks's burette; fig. 372 is Mohr's burette. In this, the most perfect form of burette, the pipette has only the additions marked *a* and *b*. The jet *a* is of glass; *b* is a metal pinchcock, crossing a caoutchouc tube, which it opens and closes at pleasure.

The variety of pipettes, burettes, and other instruments requisite for the practice of *Volumetric Analysis*, is so considerable, that we refrain from giving further details, and refer the reader to *Griffin's* 'Chemical Handicraft,' in which work the subject of Volumetric Analysis is treated at length.

370. PRICES OF A FEW VARIETIES OF BURETTES.

Gay-Lussac's form, fig. 370 }
Binks's form, fig. 371 } All at the prices named below.
Mohr's form, fig. 372 }

THE BURETTE. 87

A. Contents, 50 septems, in 50 spaces. *Price 2s. 6d.*
B.   ,,   100 septems, in 100 spaces. *Price 3s. 6d.*
C.   ,,   1000 grains, in 100 spaces. *Price 4s.*
D.   ,,   50 grammes, in 100 spaces. *Price 3s. 6d.*

E. Mahogany foot for Gay-Lussac's and Binks's burettes, same as fig. 214. *Price 1s. 6d.*

F. Mahogany support for Mohr's burette. Fig. 372. *Price 5s.*

Consult *Griffin's* 'Chemical Handicraft' for full particulars of burettes, and their prices.

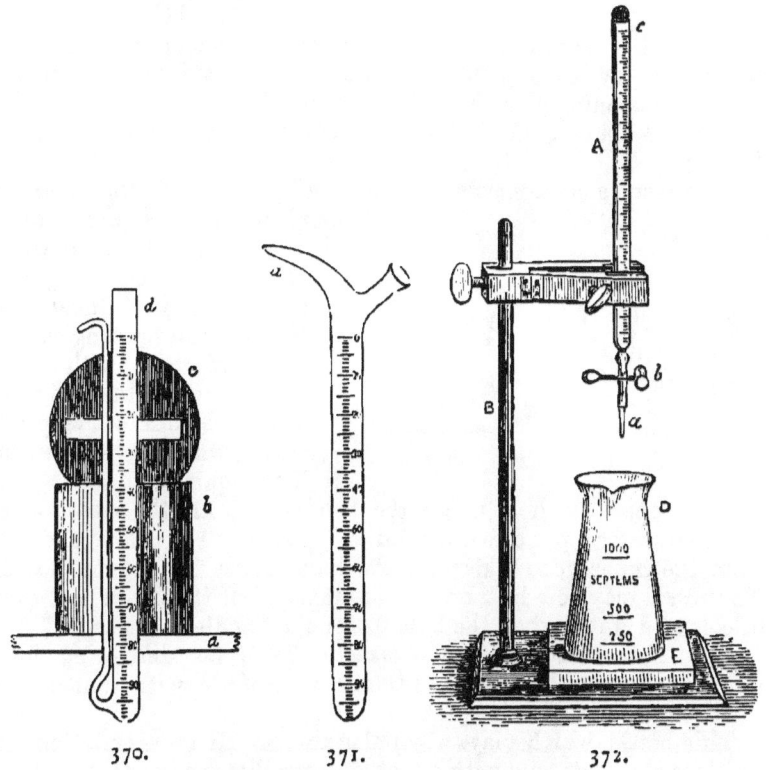

370.   371.   372.

TOYS FOUNDED ON THE PRINCIPLE OF THE PIPETTE.

373. *The Mysterious Funnel*, figs. 373 and 374; japanned tin-plate. *Price 2s.*

This instrument consists of two funnels, placed one within the other, and joined air-tight at the top. Fig. 373 represents the external appearance of the funnel; fig. 374 shows it in section. There is a small hole under the handle which communicates with the space between the two funnels.

Close the bottom of the funnel by the finger, leave open the hole under the handle, and fill the funnel with coloured water; then close the hole under the handle and open that at the bottom, upon which all the coloured water will escape from the inner funnel, but will leave the closed space full. The inner funnel may then be filled with common water, both the holes being kept closed. It is now in the operator's power to deliver the colourless water alone by opening the bottom hole, or the two kinds of water mixed, by opening also the hole under the handle.

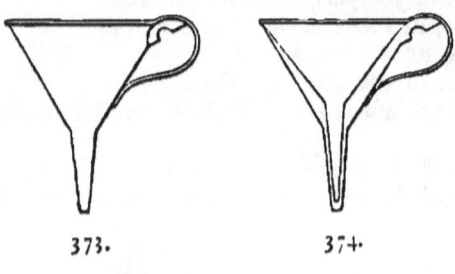

373. 374.

375. *The Magic Can*, figs. 375 and 376; japanned tin-plate. Price 3s.

This instrument consists of two vessels, one within the other, as shown by the figures, where 375 represents the outside of the apparatus, and 376 a section of it. Under the handle is an opening which communicates with the space between the two vessels, and admits the pressure of atmospheric air when it is required.

375. 376.

Leave open the hole under the handle and fill the vessel with coloured water; put down the lid and pour out the coloured water from the inner vessel, keeping the hole under the handle closed. The vessel may now be shown to be empty; but it delivers a stream of coloured water when the hole under the handle is opened.

377. *Robert Houdin's inexhaustible Bottle* for delivering four different liquors, size of an ordinary wine-bottle, with funnel and small wine-glass. *Price* 5s.

This bottle, which plays a popular part in all representations of *physique amusante*, and with which one can dispense to each of four assistants a glass of wine at his choice, is made of tin-plate and painted to imitate a common black wine-bottle. It is divided within into four compartments, representing so many pipettes, each having the upper orifice leading to a hole in the outer surface of the bottle, where it can be safely closed by a finger, and a jet which represents the neck of the bottle. The compartments being previously filled with the four liquors, the process of giving to each person the wine

THE SYPHON.

that he asks for consists in liberating the mouth of the particular pipette that contains the desired wine. The four holes in the outer surface of the bottle are necessarily so placed as to be manageable by the fingers and thumb of the operator's right hand.

381. *The Syphon.*—The syphon is a bent tube, one leg of which is longer than the other. When it is filled with a liquid, and the short leg is put into that liquid, a current runs out of the long leg until the supply of liquid sinks below the end of the short leg, when the action ceases. The two principal uses of the syphon are to transfer liquors from vessel to vessel, or to run off a clear liquid from above a sediment or a precipitate. The following figures show several varieties of the syphon :—

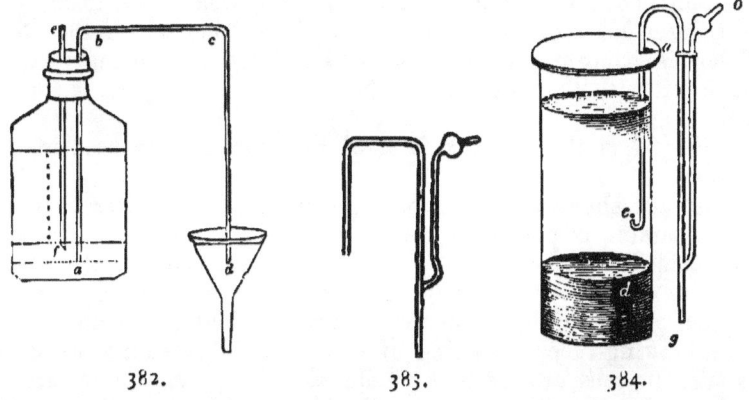

382.   383.   384.

Fig. 382, letters *a*, *b*, *c*, *d*, represent the simplest form of the syphon. As represented in this figure, it is set in action by air

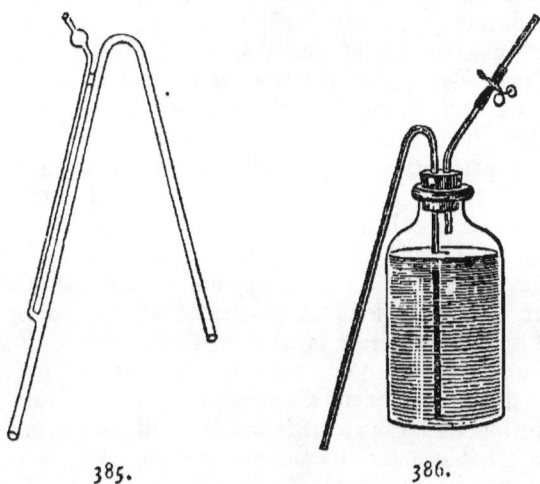

385.   386.

blown in by the tube, e, f; and, once in action, it continues to act till the liquor is all transferred, or till the tube, e, f, is stopped at e. There is an advantage in the arrangement shown in this figure as respects the operation of washing a filter, which goes on with great regularity when the water is driven over at a constant rate by the pressure of air in the tube $e f$.

Fig. 383 represents a syphon with a suction tube. Fig. 384 represents such a syphon as 383 in action. To use it, the short leg of the instrument, c, is dipped in the liquid that is to be transferred, a finger is put on the end of the long tube, g, and the liquor is to be sucked up by the side tube at o until it is seen to run over the bend at a into the long leg; the mouth at the side tube and the finger at the bottom of the long leg are then to be removed. At d in fig. 384 is a large mass of sediment or precipitate, from which the clear liquor is being syphoned. To avoid stirring up the mass d and mixing it with the clear liquor, the end, c, of the syphon is turned a little upwards.

Fig. 385 is the ordinary form of the syphon when used in the arts.

Fig. 386 shows a method of syphoning, in which every part of the apparatus is put under control, and which, therefore, affords a complete illustration of the operation. It consists of a bottle which contains the liquor that is to be drawn off; secondly, of a plain syphon passed through a stopper of cork or caoutchouc; and lastly, of a blowing-tube, consisting of two glass tubes connected by a caoutchouc tube, crossed by a pinchcock. Supposing the apparatus to be newly set together, you put the blowing-tube into your mouth, squeeze the pinchcock, and blow into the bottle, upon which the liquor passes through the syphon and escapes. If you wish that action to continue, you have only to slip the pinchcock from the caoutchouc tube to one of the two short glass tubes, when the operation proceeds; but if you wish the action to stop, you remove the pinchcock back to the caoutchouc tube and permit it to press it closely, when the action ceases.

Fig. 387 represents the application of a syphon to the decantation of a solution, without the direct access of atmospheric air. Suppose it to be a solution of potash, contained in a stock-bottle, which is to be let off by the syphon from time to time in small portions for use as a test. The syphon is fixed, as shown on the figure, with a burette jet and pinchcock at the bottom. In the stopper of the stock-bottle is placed a tube, containing lumps of potash for absorbing carbonic acid from the air that passes through it. Under this arrangement the syphon remains always full of the testing-solution, and when a little is drawn off its volume is supplied by purified air that enters by the upper tube.

Fig. 388 represents a syphon, surmounted by a stopcock, to be used instead of a suction tube to set the syphon in action.

Fig. 389, the Würtemberg syphon. This acts like a common syphon, although the two legs are of the same length. When it is to be set in action it must be filled with some of the liquor to be syphoned, and then have one leg dipped into that liquor.

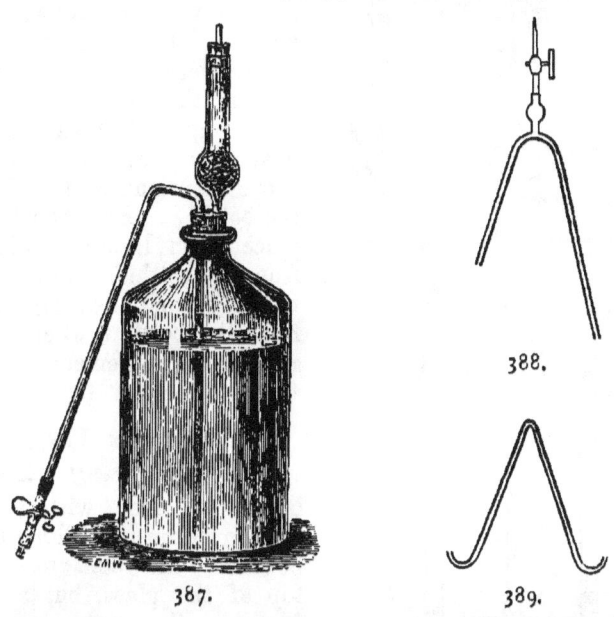

387.   388.   389.

390. Fig. 390 is an interesting example of the use of syphons in the arrangement of chemical apparatus. The figure represents a pair of Mohr's burettes, fitted to two large stock-bottles for supplying countervailing solutions for volumetric analysis. All the parts from a to g stand on the worktable; the other parts, h to l, are mounted on a shelf. The purpose is to transfer from time to time small quantities of test liquors from the two bottles, $k\ l$, to the graduated tubes, d, e. These transfers are effected as follows:—
Suppose a small quantity of test-liquor is to be passed from the bottle k into the glass beaker placed below the left-hand tube. The pinchcock f is opened, upon which the liquor passes from the tube d into the beaker. The quantity is measured, not only by the graduation on the vertical tubes, but by a circular line on the floats d and e, which serve to check the other graduations. When this small quantity of liquor passes out of the tube d, its place is immediately supplied by air, which enters from i by the syphon g, and afterwards the tube d is supplied with other liquor which comes by the syphon h from the bottle k; but when this supply of liquor takes place, some

air must go out of the tube *d* to make room for it. That air goes up by the tube *g* into the bottle *k*. But since a certain mass of liquor passes out of the system of apparatus into the glass beaker, an equal volume of air must come into it to preserve the equilibrium. That equivalent volume of air comes in by the tube *i*, in passing through which tube the air is deprived of carbonic acid, dust and other matters that might prove injurious to the liquor in the bottle *k*. Exactly what takes place when liquor is abstracted from the bottle *k*, takes place also when a similar abstraction is made from the bottle *l*, where there is an equal arrangement of syphons and air tubes.

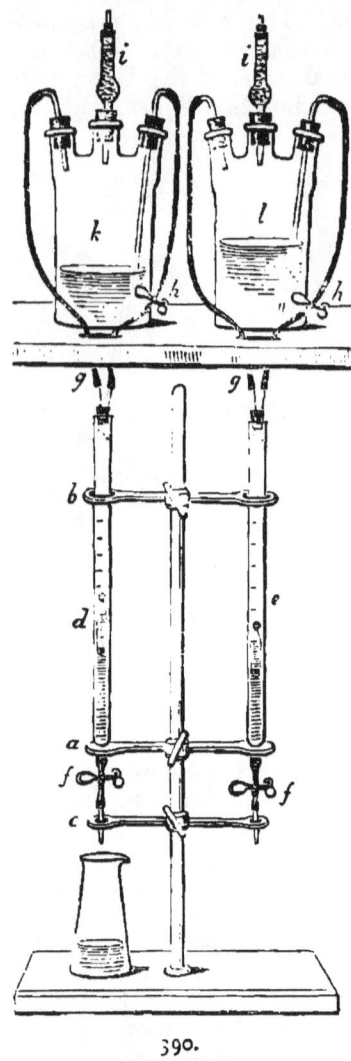

390.

SYPHON TOYS.

391. *Tantalus's Cup.*—A glass beaker on foot, with a curved syphon within it. This glass can be filled with water *nearly* to the top of the glass, but *not quite*; for when the water rises to the upper part of the bent syphon, the latter begins to act, and the water all runs out. The stem and foot of the apparatus are both perforated, and the lower end of the syphon is fixed there by a cork. Fig. 391. *Price* 1s. 6d.

A short tube passes from the cork *downwards*, to carry off the liquid.

392. *Hempel's Syphon*, another form of *Tantalus's Cup.*—Nearly of the same construction as No. 391. The central tube, open at both ends, is part of the outlet tube; the tube that covers it is put over it loosely. When the water rises to the top of these tubes, it flows out by the syphon action. Fig. 392. *Price* 1s. 6d.

394. Sometimes Tantalus's cup is made of glass or japanned tin-plate, having within it a figure of king Tantalus, within which a

PRICES OF SYPHONS.

syphon is concealed. Water is poured into the vessel until it nearly reaches the lips of the figure, whereupon the syphon begins to act and draws off the whole of the water. *The price of this toy is about* 10s.

395. *Fountain Syphon*, or Eye Fountain, for bathing inflamed eyes, fig. 395, 24 inch. *Price 6d.*

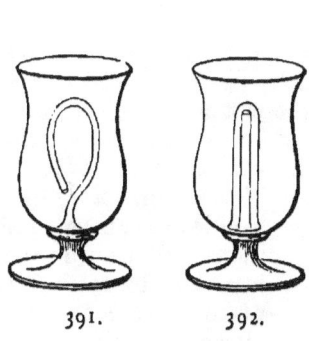

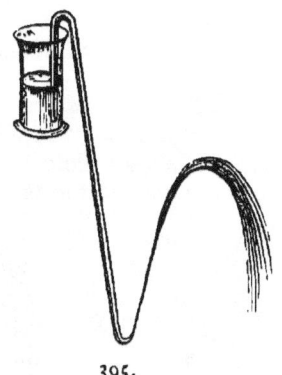

391.   392.   395.

Prices of Syphons.

382. Gay-Lussac's plain syphon, 24 inch. *Price* 1s.

382 A. Gay-Lussac's plain syphon, fitted with regulating-bottle as represented by fig. 382. *Price* 3s. 6d.

384. Syphon with suction tube, 12 inch. *Price* 9d.
      24 inch. *Price* 1s. 3d.

385. Syphon with suction tube, common form, 36 inch. *Price* 1s. 6d.

386. Syphon apparatus, with blowing-tube, fitted for explaining the theory of the syphon. *Price* 4s.

387. Syphon fitted to bottle for test liquor. *Price* 3s.

388. Syphon with stopcock. *Price* 3s. 6d.

389. Würtemberg syphon, 36 inch. *Price* 1s. 6d.

Intermitting springs, described in another section, depend upon syphons for their action.

Syringes.

403. *Glass Syringe*, about 8 inches long, 1 inch diameter, fig. 403. *Price* 1s. 6d.

404. *Glass Syringe*, about 8 inches long, 1 inch diameter, with bent point, fig. 404. *Price* 2s.

403.

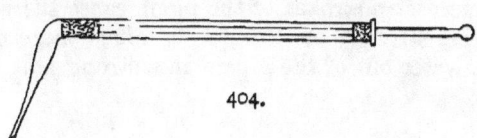

404.

Glass syringes of smaller sizes, namely:—
405. 3 inches long. *Price* 4d.
406. 4 inches long. *Price* 6d.
407. 5 inches long. *Price* 8d.

The pistons of all these glass syringes are solid and covered with cotton wool. The piston rods are guided by corks fixed in the ends of the tubes. There are no metal fittings.

GLASS WORKING MODELS OF PUMPS.

412. *Lift Pump, a working model in glass*, with glass valves; size of barrel about 10 inches by 1¼ inch; fig. 412. *Price* 4s. 6d.

Figure d shows the construction of the piston c, which contains a valve that opens upwards. The valve at b is closed and weighted with mercury. When the piston is drawn up by the handle at e, the valve at b rises and admits water into the barrel of the pump. When the piston is depressed the valve in d permits the water to rise towards a, and the next stroke expels it from the spout of the pump.

The water pan is not included in the price of the pump. A piece of caoutchouc tube can be put on the neck b, and the pump can be mounted on a jar like the pump in fig. 425.

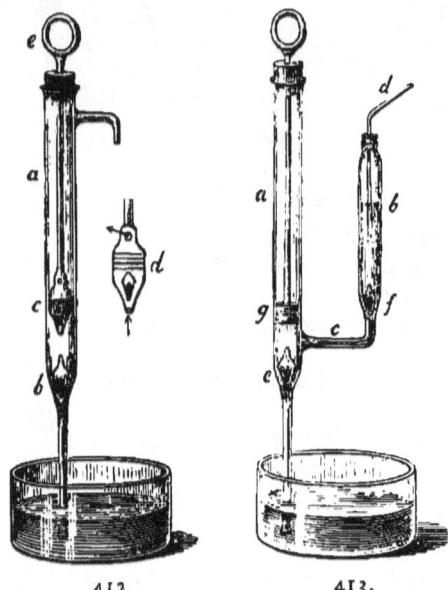

412.   413.

413. *Force Pump, a working model in glass*, with glass valves: size of the barrel about 10 inches by 1¼ inch; fig. 413. *Price without the glass pan*, 4s. 6d.

The valves at e and g are both solid. When the piston is drawn up, the valve at e rises and admits water into the barrel a. When the piston is pressed down the valve e closes, and the water is forced through the pipe c, and past the valve f into the air-chamber b. Successive strokes of the pump cause the water in the air vessel to press strongly on the air contained therein, which in turn forces the water out of the pipe d in a strong jet.

WORKING MODELS OF PUMPS.

414. *Force Pump, a working model,* with large air-vessel similar to that of a fire-engine, made in glass, with coloured glass valves; mounted on a mahogany frame; fig. 414. Price *without the basin,* 8s.

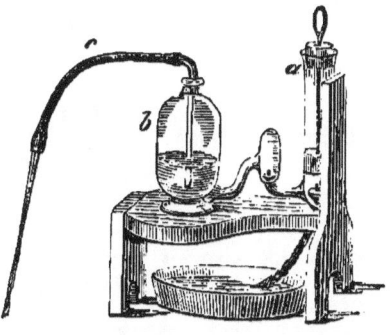

414.

a is the force pump, *b* the air vessel, *c* a flexible tube carrying the jet pipe.

415. *Fire Engine,* or double-barrelled force pump, a working model, made in glass, with coloured glass valves, mounted on a mahogany frame; fig. 415. *Price without water basin,* 12s. 6d.

a, a, are the two force pumps; *b* is the air vessel; and *c,* a flexible tube carrying the delivery pipe or jet.

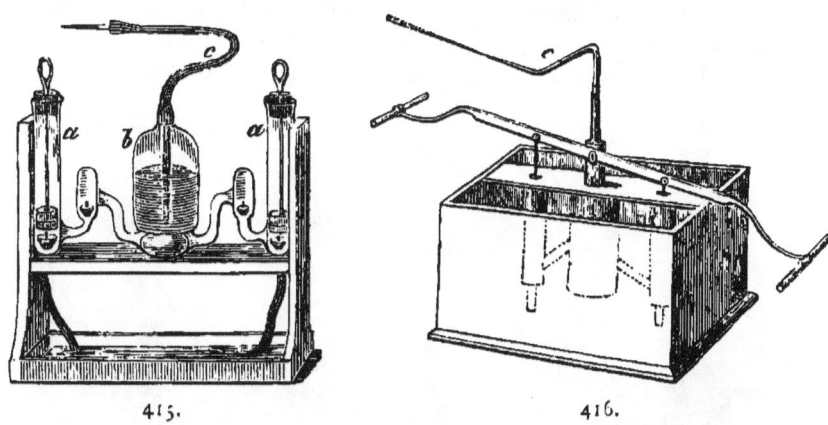

415.   416.

416. *Fire Engine,* a working model of a fire engine, or double-barrelled force pump, japanned tin-plate. Price 7s.

This model will throw a jet of water to a distance of 20 feet or more; but it cannot be opened to show the interior construction. The glass models are better adapted to explain the mode of action of a fire-engine.

WORKING MODELS OF PUMPS, IN GLASS, MOUNTED WITH BRASS.

418. *Valves.*—Models of 5 varieties, namely, the butterfly valve, bellows valve, round spring valve, conical valve, and oil-silk valve, in stained hard wood, each 6 inches in diameter. *Price of the set,* 10s. 6d.

HYDRODYNAMICS.

Valves are a species of flood-gates, that open a passage for a stream of liquid or gas in one direction, and close the passage against the stream in the contrary direction. These valves are models on a large scale of valves that are used for pump work.

a. Butterfly valve.

d. Conical valve.

b. Bellows valve.

e. Oil-silk valve.

c. Round spring valve.

421. *Lift Pump*, a working model, of a superior description, with stout glass body, about 7 inches long by 1¼ inch wide, mounted with brass; fig. 421. *Price without stand*, 16s.

422. *Force Pump*, a working model, of a superior description,

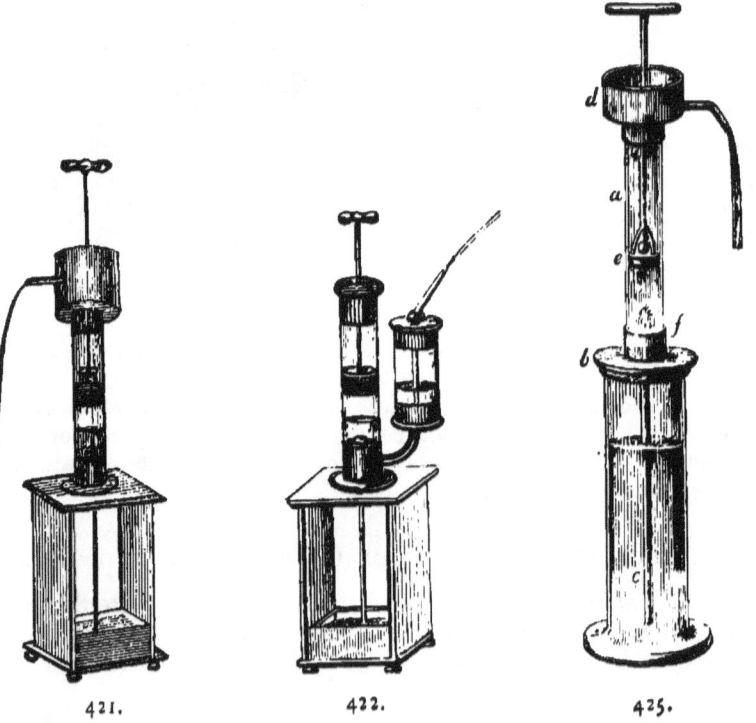

421.   422.   425.

with stout glass barrel, about 7 inches long by 1¼ inch wide, mounted with brass; fig. 422. *Price without stand,* 25*s.*

423. *Pump Stand,* a polished mahogany frame with a japanned tin-plate water cistern, to suit either of the pumps, Nos. 421 and 422, as represented in these figures. *Price* 9*s.*

425. *Tate's School Lift Pump, a working model, large but cheap,* with a glass barrel about 10 inches long, and 2 inches diameter, mounted in japanned tin-plate, with a wide trough at top, and stone marbles for valves. *Price* 7*s.*

425 A. Glass Cylinder to contain water to supply this pump when in use; size 14 inches high, 4 inches diameter, with a board for the pump to rest upon. *Price* 3*s.* 6*d.*

This apparatus gives a continuous stream of water, and the action is visible by a Class. Before it is exhibited it must be examined, to see that the cotton packing of the pistons is in good condition. At first, some water must be poured into the trough *d,* to wet the pistons and enable them to commence action.

427. *Water rises in Pumps in consequence of the pressure of air on the Water in the Reservoir.*—The apparatus for this demonstration is represented by fig. 427.

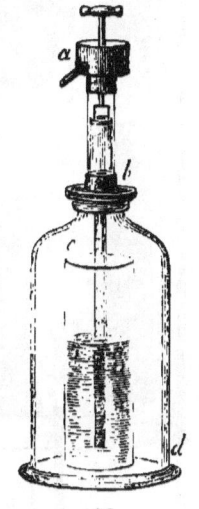

The part *a, b,* is the model of the working part of a common lift pump, see No. 421. At the lower part of this is a ground brass plate, to rest upon the ground neck of the receiver *c, d.* To the centre of this plate a tube can be screwed, which descends into the water jar placed under the receiver. When the air-pump is first worked, before the air is exhausted from the receiver, the water flows freely from the spout; but, when the exhaustion of air is effected, the pump ceases to deliver water. When air is admitted into the receiver, the pump resumes its power to deliver water.

Price of this apparatus, as represented by fig. 427, 1*l.*

SPRINGS AND FOUNTAINS.

There exist a great variety of Springs and Fountains. Our apparatus afford illustrations of the following six varieties, viz.:—

439-445. Fountain produced when water falls from a high reservoir and supplies a jet.

446-459. When condensed air forces a water jet into free air. Condensed air fountains.

HYDRODYNAMICS.

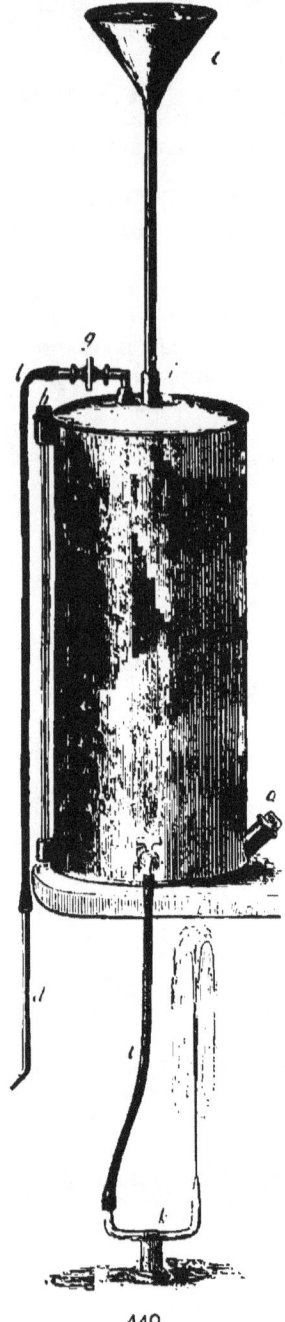

440.

460-470. When uncondensed air forces a water jet into an exhausted receiver. Fountains in vacuo. Heron's Ball.

471-475. When a column of water condenses a confined portion of air and causes it to force a water jet into free air. Heron's Fountain.

476. When a fall of water from a syphon creates a vacuum into which a water jet is forced by free air.

480-482. When an interrupted supply of air causes a spring to be intermittent.

439. Fountains produced when water falls from a high reservoir and supplies a jet. This is the ordinary method of supplying fountains for public gardens such as that of the Crystal Palace. The water is pumped up to a high reservoir and then allowed to descend through a system of pipes. With the following apparatus the experiment can be shown in a lecture room.

440. *Fountain produced by a Fall of Water from an eminence under common atmospheric pressure.*—Fig. 440 represents an ordinary *Chemical Gas Holder*. It is of a cylindrical form, and is made of japanned zinc; the body is 18 inches high, and $10\frac{1}{2}$ inches diameter; contents 1500 cubic inches. It is represented as standing on a table or shelf, and in use for hydraulic experiments. We shall, however, explain the references: a is the neck by which gas can be delivered into the gas holder, the 2 stopcocks being closed at the time, upon which water runs out of the neck a, bulk for bulk, with the gas that is passed in. c is a funnel by which water is put into the reservoir. It is made 18 inches long, to give a pressure when required. $g\ l\ d$ show the means of passing out gas in any direction. At h is a wide gauge-pipe, to show the level of the water in the gas holder at any time. The contents are shown by a graduation into spaces of 50 cubic inches.

SPRINGS AND FOUNTAINS. 99

The fittings of the gauge pipe are so constructed that, if the glass tube should happen to be broken, it can be easily replaced by another, the screw *h* being moveable.

440. Price complete, as represented by fig. 440, 2*l*.

441. Price of the gas holder without the lower stopcock and the pieces marked *i*, *k*, 1*l*. 10*s*.

442. Price of the lower stopcock, 4*s*.

443. Price of the pieces *i*, *k*, viz., fountain jet and 6 feet of caoutchouc tube ¾ inch bore, 6*s*.

444. *Water Bottle*, by which a supply of water can be provided for a fountain on a small scale. The bottle, filled with water, is to be placed on a high shelf, and the tube that is to supply the jet is to be fixed on the stopcock *c*.

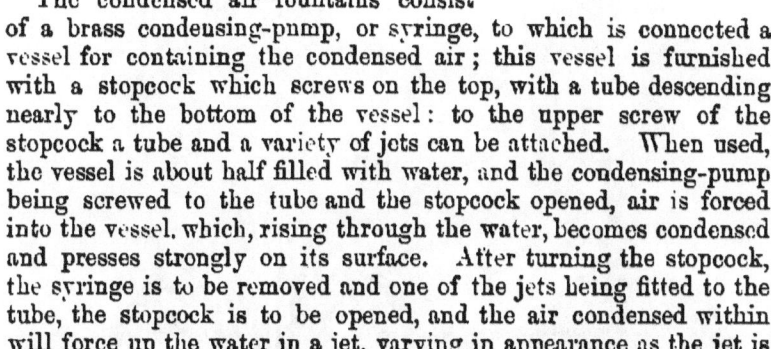

444.

444. Price of two-gallon jar with brass cock, 7*s*.

445. Price of five-gallon jar with brass cock, 15*s*.

446. Fountains produced when condensed air forces a water jet into free air.

There are three varieties of this kind of fountain, Nos. 447, 448, 449.

The condensed air fountains consist of a brass condensing-pump, or syringe, to which is connected a vessel for containing the condensed air; this vessel is furnished with a stopcock which screws on the top, with a tube descending nearly to the bottom of the vessel: to the upper screw of the stopcock a tube and a variety of jets can be attached. When used, the vessel is about half filled with water, and the condensing-pump being screwed to the tube and the stopcock opened, air is forced into the vessel, which, rising through the water, becomes condensed and presses strongly on its surface. After turning the stopcock, the syringe is to be removed and one of the jets being fitted to the tube, the stopcock is to be opened, and the air condensed within will force up the water in a jet, varying in appearance as the jet is varied.

447. *Condensed Air Fountain*, consisting of a glass globe on foot, in 1 piece, stout Bohemian glass, 6 inches in diameter, and nearly ¼ inch thick in glass, with brass collar, water tube and stopcock. Fig. 447, which shows this fountain in action, when mounted with jet No. 447. *Price of glass fountain and stopcock, without jets*, 31*s*. 6*d*.

447 A. The fountain, with the 4 jets complete. *Price* 52*s*. 6*d*.

448. *Condensed Air Fountain*, large size, made of stout sheet

H 2

HYDRODYNAMICS.

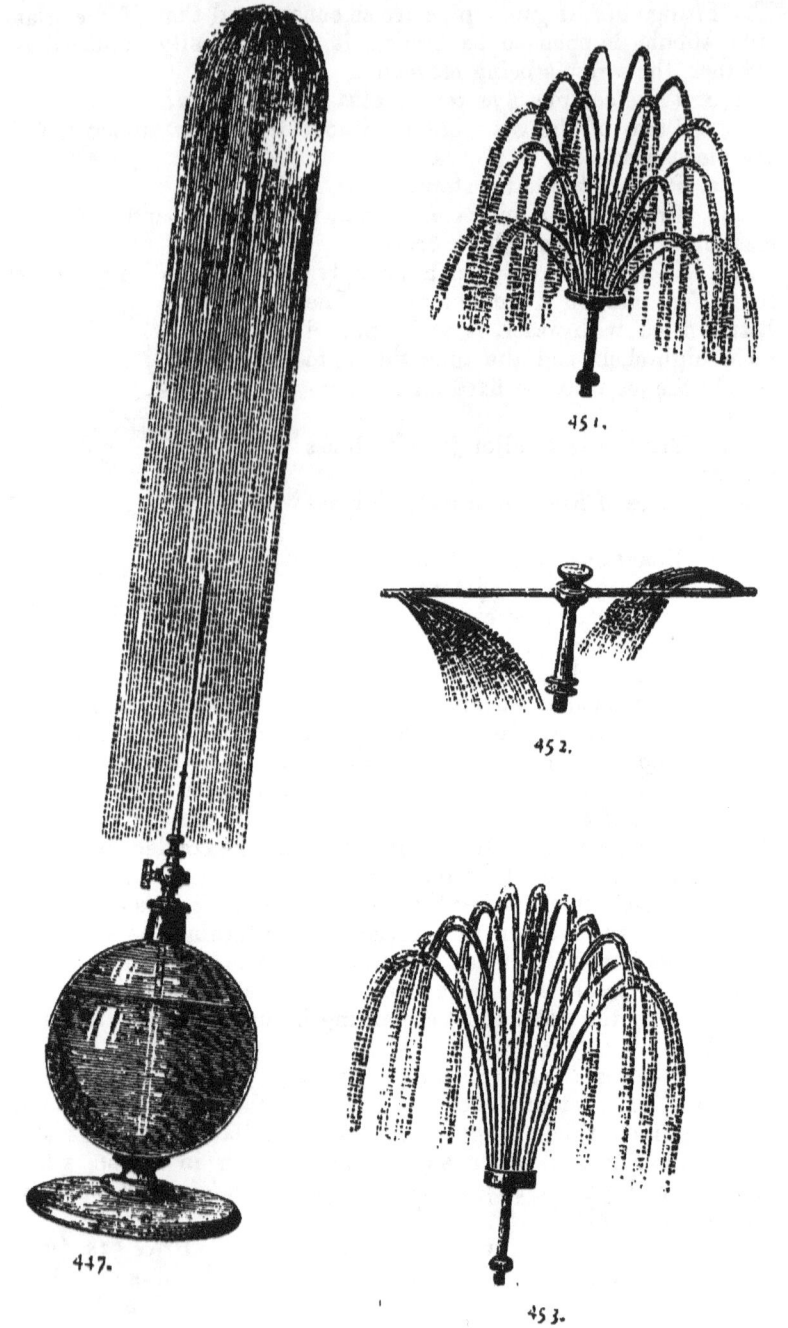

447.

451.

452.

453.

zinc, or zinced iron; size 12 inches in height, 6½ inches in diameter, with extra neck, 8 inch water basin, and two stopcocks; fig. 448. *Price, without jets or syringe, 42s.*

The use of the extra neck is to permit the supply of additional air while the fountain is in action.

448 A. The fountain with the 4 jets complete. *Price 63s.*

449. *Condensed Air Fountain,* small size, brass cylinder, measuring 6¼ inches in height, 2½ inches in diameter; fig. 449. *Price, with a stopcock but without jets,* 12s.

449 A. The fountain with the 4 jets complete. *Price 31s. 6d.*

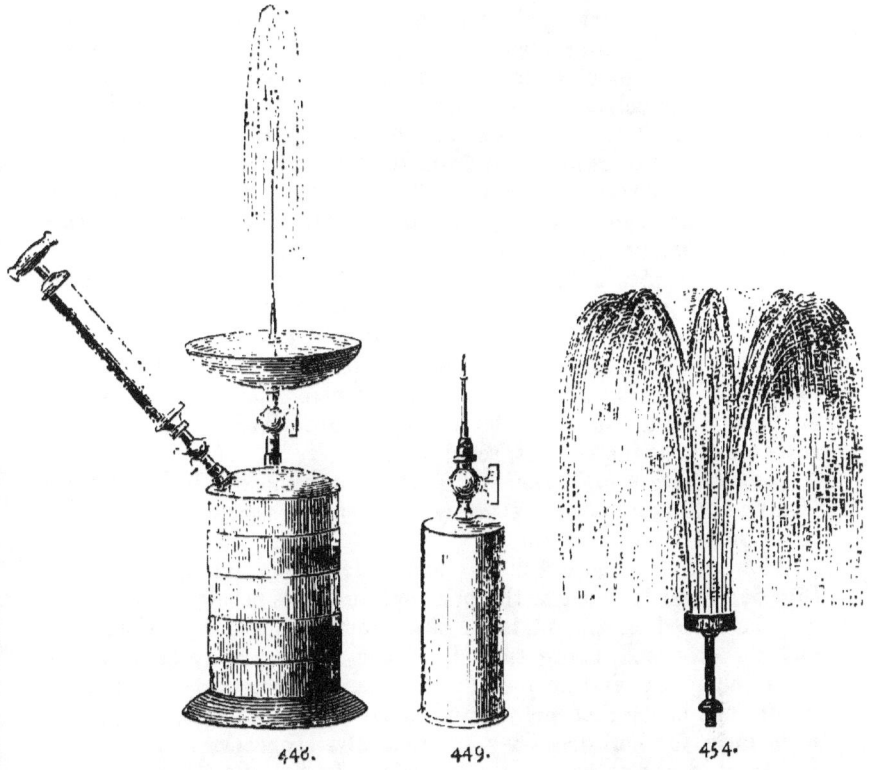

448.   449.   454.

450. *Fountain Jets,* a set of 4, figs. 447, 451, 452, 453. They suit all the 3 fountains, and are sold either as a set or separately. Price of the set, 22s.

Prices separately:—

	s.	d.		s.	d.
447.	2	0	452.	9	6
451.	5	0	453.	5	6

The jet, No. 454, is shown by jet No. 453, when under strong pressure. No. 452 is a revolving jet.

Condensing Syringe for supplying air to the fountains, fig. 455.

455. Small size. *Price* 8s.
456. Large size. *Price* 16s.
457. *The Chromatic Fire-cloud.* —The chromatic fire-cloud is produced when the air vessel is supplied with a strong solution of chloride of strontium and chloride of copper in spirit of wine, instead of water. Air is condensed into it, the stopcock closed, the syringe removed, and a jet applied as above described. On the stopcock being opened the compressed fluid is driven out with great violence, and must be projected on a wall or the ceiling of a darkened room; a light should be applied to the vapour on the ceiling, and small quantities of the liquid must from time to time be projected so as to sustain the flame. The effect produced is exceedingly brilliant, as the luminous waves are of almost every colour.

460. Fountains produced when uncondensed air forces a water jet into an exhausted receiver.—*Fountains in vacuo.* Three varieties, Nos. 461 to 463.

461. The apparatus represented by fig. 461 shows a method of making such a fountain by using the table of Tate's air-pump as a transfer plate. It is described in the article respecting that instrument, see § 516.

462. *Fountain in Vacuo.*—Fig. 462 shows the *Single Transferrer* (fig. 627) in active use. *Price of transfer plate on iron foot without basin and receiver*, 17s. 6d.

The transfer plate can be unscrewed from its wooden base (fig. 627) and be fixed to the pump plate by its screw. A receiver may be placed on the plate of the transferrer, and be exhausted; and the stopcock, being turned off, the receiver may be removed from the pump, and be placed on a suitable support. If then a water tube is attached below, and a jar or pan of water supplied, it is only further necessary to open the stopcock to produce a fountain in the receiver; see fig. 461. In fig. 462 the transferror is supported by a special foot screwed on the stopcock.

463. *Fountain in vacuo, without a transfer plate.* - Represented by fig. 463. The glass receiver is mounted with a collar, a stopcock, and a fountain jet. It is exhausted in the usual way, then removed from the air-pump, screwed upon a foot, and placed in a water pan. The stopcock being then opened, the jet is formed in the receiver. If desired, the stopcock can be screwed to a water

SPRINGS AND FOUNTAINS. 103

tube, and set on a water bottle like fig. 461. *Price of fountain on foot with receiver, but without the water basin, 24s.*

461.   462.   463.

467. *Heron's Ball*, to produce a fountain in vacuo.—This apparatus consists of a glass flask one-half full of water, into which a glass tube, dipping into the liquor, is securely fastened by a cork. When this flask is placed under an air-pump receiver, about to be exhausted, the subsequent expansion of the air within the flask forces the water through the tube, forming a fountain in vacuo. If air is admitted into the receiver, the effect instantly ceases.

A. Size of flask, 3 ounces of water. *Price 1s.*
B. Size of flask, 6 ounces of water. *Price 1s.*

Any glass receiver sufficiently wide and tall may be used for this experiment. The moveable pump plate (No. 563) may be used, with a view to keep the water from the pump itself.

468. *Heron's Ball*, a more elegant form, globular glass on foot,

with brass fittings; fig. 468. *Price without the receiver and glass basin*, 6s.

471. Fountains produced when a column of water condenses a confined portion of air and causes it to force a water jet into free air.—*Heron's Fountain.*

472. Heron's *Fountain*, for showing the elastic force of compressed air; the middle tube and half of each ball should contain air; the rest be filled with water, including the funnel.

a. Price, mounted on a mahogany stand like *fig.* 472, *but without the glass cistern*, 6s.

b. Price of the *glass fountain not mounted*, 3s.

473. Heron's *Fountain*, another pattern, and style of mounting, as shown by fig. 473. *Price with wooden foot*, 6s.

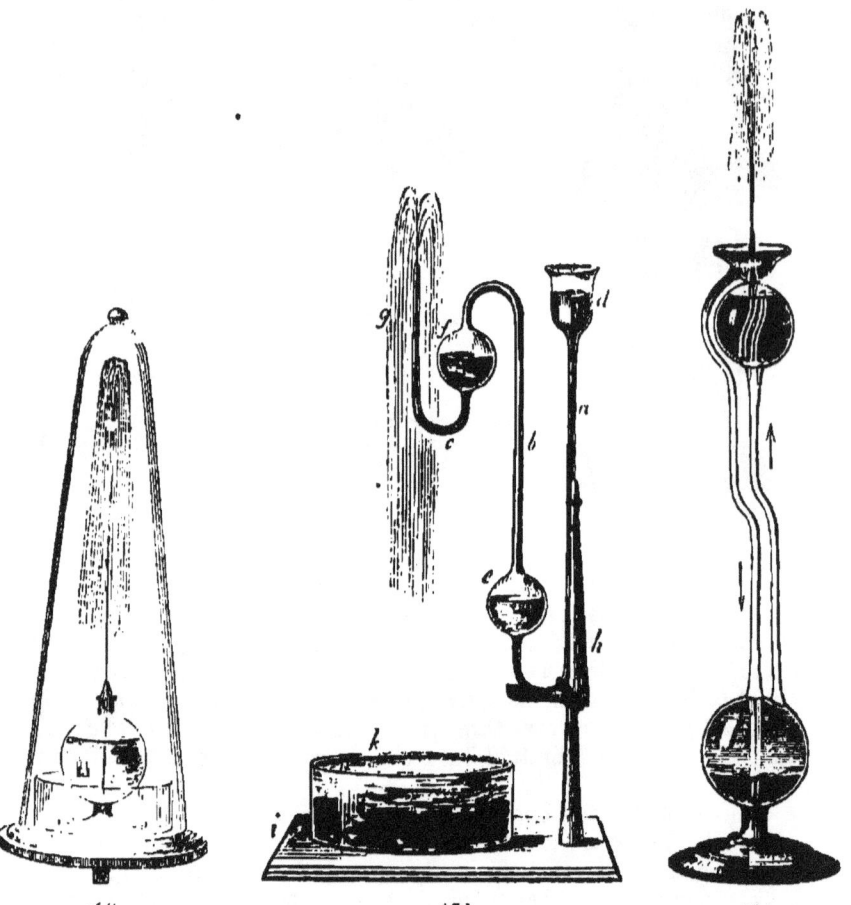

468.   472.   473.

The fountains (figs. 472, 473) are charged with water by partly filling them at the top, and then turning them upside down for a short time. When they are properly charged, the action is kept up by putting water into the funnel.

Care must be taken to use water free from bits of straw, or other material that can stop up the jets.

474. *Heron's Fountain, made of japanned tin-plate,* height about 15 inches; figs. 474, 475. *Price* 4s.

Atmospheric air is capable of being condensed or reduced in volume to a very considerable degree; but, when it is thus compressed, it has a tendency to recover the same volume or space which it originally occupied, and hence exerts a pressure on the objects by which it is enclosed.

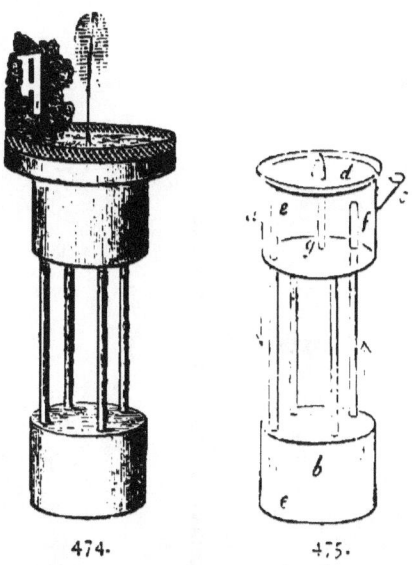

474.   475.

The action of Heron's Fountain depends mainly upon this principle. The upper reservoir a being first filled with water at the opening c, which must then be securely stopped by a cork, water is poured into the basin d at the top of the reservoir. This water runs through the tube e into the lower reservoir b, which had previously contained nothing but air. The air, being compressed by the entrance of the water, forces its way up the tube f into the upper reservoir; here, finding no vent, it acts like a spring on the surface of the water in the reservoir, and forces the liquid out through the jet tube $g\ l$, thus producing the very pleasing effect of a miniature fountain.

When the fountain ceases playing, the water may be drawn off by withdrawing the cork from the opening in the lower reservoir; but at all other times it must be kept carefully closed. The other columns have no connexion with the interior, and are merely intended as supports. This fountain derives its name from its inventor, Heron, of Alexandria.

476. Fountain produced when a fall of water from a syphon creates a vacuum into which a water jet is forced by free air.

477. *Fountain produced by the action of a syphon.*
Price of the syphon, consisting of 3 *glass tubes and a cork,* 4s. 6d.

Price of the iron support, 9s.
Price of the glass beaker, 1s.

The pieces of this apparatus being put together as represented in fig. 477, and the beaker being filled with water, the fountain is set in action by sucking air from the lower end of the long tube. The pressure of the atmosphere upon the water in the beaker then forces a jet up into the fountain glass. The proper play of the apparatus depends upon the proportions which the orifices of the two tubes bear to one another.

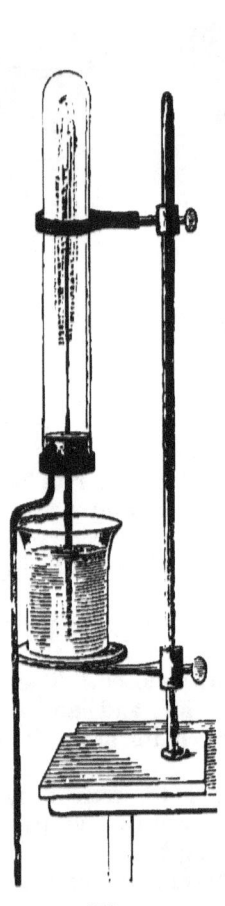

477.

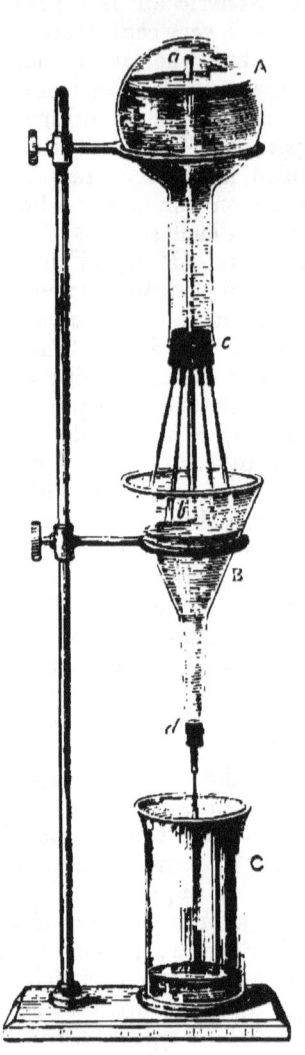

481.

SPRINGS AND FOUNTAINS.

480. *Intermittent springs.*
481. *Glass model of an intermitting spring* (fig. 481).

	s.	d.
Price of the flask and funnel A B, with the tubes attached	5	0
The glass receiver, c	2	0
The tall iron stand	9	0
The set complete	16	0

Use.—The stand, the receiver C, and the funnel are first arranged; the flask A is then nearly filled with water, and the cork with tubes is adjusted at c. The tube a goes into the funnel a little more than halfway down. Some water is put into the funnel to cover the lower end of the tube a. The whole is then in working order, providing the tubes are all clean and clear. *The action is as follows:* When the pipe d has carried off some of the water from the funnel, so as to uncover the lower end of the tube a, air passes through that tube up into the globe A, whereupon the water in the globe passes through the four narrow delivery tubes into the funnel, in which it accumulates so as to stop the passage of air up the tube a. The action of the fountain then ceases, until the outlet tube d has again liberated the end of the tube a; and thus an intermitting action is sustained, until the water in the globe A is exhausted. It is of importance to regulate the flow from the jet d. If it is either too slow or too rapid, the action ceases. But it is easy to keep it in order. All the pipes must be kept clean.

482. *Glass model of an Intermitting Fountain.*—The description of the Intermitting Spring (No. 481) applies pretty well to this apparatus also. A fall of water from the lower mouth of the water bottle causes a fountain to rise in the funnel. The water speedily rises in the funnel and closes the lower end of the long tube that carries air to the upper part of the water bottle. When the supply of air is thus cut off, the water ceases to run out of the bottle and the fountain ceases to play. But in the meantime the water flows from the funnel into the glass pan, and sets free

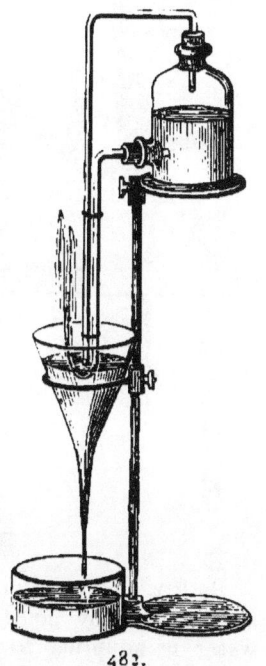

482.

the air tube in the funnel, and the action of the apparatus depends a good deal upon the regularity of that flow of water from the funnel. The height of the fountain in the funnel depends upon the length of the two tubes that separate the funnel from the bottle: the longer these tubes, the higher is the jet.

		s.	d.
Price of the fountain without the stand	.. ..	6	0
" iron stand	.. .. ..	9	0
" water basin, 6 inches diameter		1	3
Complete		16	6

WATERWORKS.

486. *Water-wheels*, a set of working models, made of tin-plate, japanned green, and consisting of an over-shot wheel, an under-shot wheel, both wheels being 8½ inches in diameter; a water cistern, a frame to support the whole in acting position, and a pan to receive the spent water.

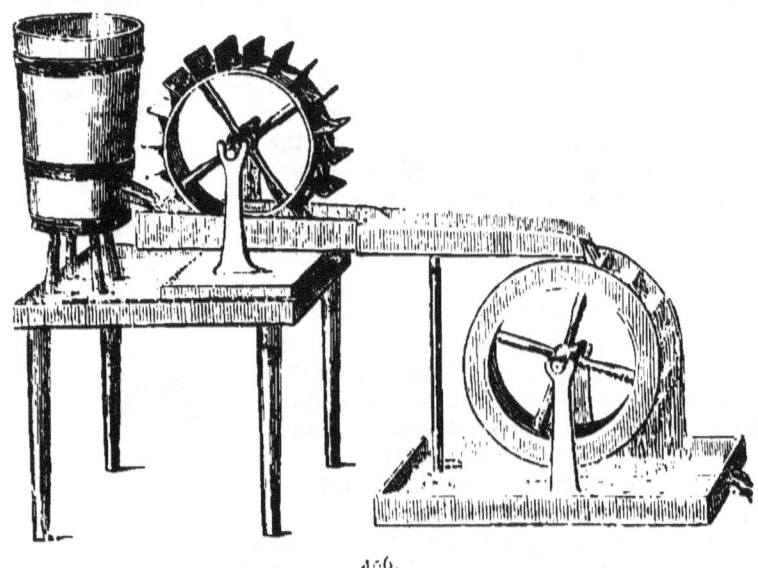

486.

Both wheels work at the same time. The water lead must be so adjusted, by gently bending the gutters, that not too much water is supplied to the over-shot wheel. Fig. 486 shows the two wheels in action. Fig. 487 shows the mode of adjusting one of

WATERWORKS. 109

the wheels, so as to make it act as a breast wheel. *Price of the set in a box,* 31s. 6d.

487.

488. *Archimedian Screw,* a working model consisting of a metal pipe wound round a cylinder, mounted on a metal frame over a water cistern, japanned. *Price* 10s. 6d.

489. *Glass Archimedian Screw,* mounted on a japanned metal frame, more elegant than the above. *Price* 21s.

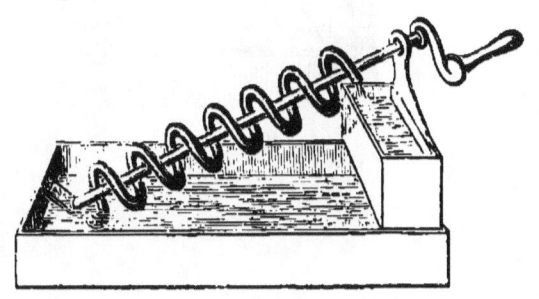
489.

Before using the screw, at a lecture, a small marble or glass ball, half an inch in diameter, should be passed through it, from below upwards, to prove the passage to be clear.

490. *Appold's Centrifugal Pump,* a working model, which gives a continuous stream of water; the acting centrifugal wheel is 1⅝ inch in diameter; it is accompanied by a separate model of the wheel 6 inches in diameter, which is one-half of the diameter of the acting wheel of the large

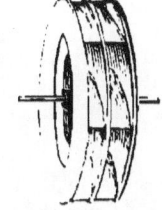
491.

110 HYDRODYNAMICS.

pump that was shown at the Great Exhibition of 1851. Fig. 491. *Price of the set*, 4*l*.

"The most advantageous application of Appold's pump is in raising large quantities of water to small altitudes; for example, it has been very successfully employed in draining fens, &c."— *Brooke*.

490.

PNEUMATICS.

AIR-PUMPS.

501. *Air-pumps of the most improved forms, and guaranteed to be in perfect action.*

All the screws of pneumatic apparatus are made of the *same size and thread*, so that the several pieces are easily fitted to one another.

If an air-pump is not used for a considerable time, the various parts require the addition of a little oil, which may be easily applied by pouring a teaspoonful into the centre hole in the brass plate, as shown at *a*, fig. 503; when a few strokes of the piston up and down will convey the oil to all the internal parts, and the machine will be in good working condition. The ground edges of all receivers should be smeared with tallow prior to being fixed on the air-pump plate. See 639. Stopcocks should be always laid aside *open*; and when a pump is put aside the blank nut, letter *s*, fig. 505, should always be screwed into the plate, to prevent the entry of dust.

502. *Exhausting power of the Air-pumps described in this section.*—I have added to the description of each pump a reference to its exhausting power. It is proper to state in what manner these powers were ascertained. This was by trial with syphon gauges. The pumps were new, and therefore in good order; the joints screwed tight together; the washers, the inside of the pumps, and the leading tubes were well supplied with oil; the receivers were carefully ground on the edges and greased with tallow; and the mercury in the syphon gauges, which were of the form represented in fig. 577, and of which several were used, had been recently boiled in the tubes. The temperature of the room in which the trials took place was about 55° Fahr., and the barometer stood at 30 inches. The results are quoted in the table (see page 112).

The *Residues* of air left in the receivers, according to these indications, are as follow:—

$\frac{1}{20}$ inch = 1 in 600	$\frac{1}{8}$ inch = 1 in 240	$\frac{1}{4}$ inch = 1 in 120
$\frac{1}{10}$ inch = 1 in 300	$\frac{3}{16}$ inch = 1 in 160	$\frac{3}{8}$ inch = 1 in 80

The kind of Pump tried. Under columns 2 to 5 is shown the atmospheric pressure indicated by the syphon gauges at the point of greatest exhaustion.	1.	2. Capacity of Receiver, 960 cubic inches; its base, 9 inches diameter.	3. Capacity of Receiver, 280 cubic inches; its base, 7 inches diameter.	4. Capacity of Receiver, 86 cubic inches; its base, 4 inches diameter.	5. Capacity of Receiver, 25 cubic inches; its base, 2 inches diameter.
		inch.	inch.	inch.	inch.
503. 2-barrel pump, 8-inch plate			$\frac{1}{4}$	$\frac{1}{16}$	$\frac{1}{8}$
504. 2-barrel pump, 10-inch plate		$\frac{3}{8}$		$\frac{1}{4}$	$\frac{1}{8}$
505. Tate's pump, 7-inch plate			$\frac{1}{5}$	$\frac{1}{10}$	$\frac{1}{20}$
519. Tate's large 10-inch plate		$\frac{1}{8}$		$\frac{1}{10}$	$\frac{1}{15}$
520. 3-barrel pump, 10-inch plate, when the vertical barrels were used		$\frac{1}{4}$		$\frac{1}{4}$	..
520. 3-barrel pump, 10-inch plate, when the horizontal pump was used		$\frac{1}{10}$		$\frac{1}{8}$	$\frac{1}{20}$
521. Pump with fly-wheel		$\frac{1}{8}$		$\frac{1}{5}$	$\frac{1}{10}$

Powers of Syphon Gauges.—For the sake of those who may wish to try the power of their pumps in this manner, I may add to the conditions of trial above cited a caution respecting the syphon gauges. In their ordinary condition—that is to say, after having been for some time exposed to the air—the syphon gauges are of no use for such trials as those recorded above. For such experiments the mercury must be *recently boiled in the gauges*, and then they retain the power of giving accurate results for only a short time. If they are exposed for two or three days to the air, they lose their power. If on the day they are made they are put under a receiver, and repeatedly exhausted and refilled with air, they are thus deprived of their proper power. The air in the gauge gets among the mercury, and puts an end to its accurate indications. When the exhausting power of a pump is tested by a gauge thus deteriorated, the exhaustion appears to be much greater than it actually is, even when the quantity of air in the gauge is so small that it cannot be seen as a bubble in the mercury, either by the naked eye or with a lens; while the existence in the gauge of a very small *visible* bubble of air will enable the pump to reduce the mercury in the closed limb of the gauge lower than that in the open limb: in other words, the exhaustion will appear to take out of the receiver more air than it contained, and thus reduce the atmospheric pressure to less than nothing.

Just as it is possible to make the power of a pump appear to be better than it is by using a fallacious gauge, so it is possible to err the other way, and by neglecting to take the precautions which I have pointed out, make the power appear to be *worse* than it is. A pump cannot, indeed, be always *new*; but it is always possible to see that its pistons and valves are in good order, that the pump is clean and well oiled, that its parts are screwed tight together and fixed firmly to a table, and that the receivers are ground smooth on the edges, and are clean and properly greased. Without taking these precautions, a pump cannot be made to work well.

Notwithstanding what is said above, the syphon gauge is highly useful for comparative experiments, and to indicate results approximately.

I have not stated in the trials recorded above the number of strokes that are required to produce each effect: that indication of power is much subject to variations from accidental circumstances, such as the greater or lesser rapidity of the strokes, and the greater or lesser accuracy with which the piston of Tate's pump is driven or pulled home to the end of the barrel at each stroke.

Neither have I answered a question that is frequently asked—How many minutes will it require with a given pump to freeze water? Generally, I may say, that any pump will freeze water over sulphuric acid, if it will produce pretty readily a vacuum indicated by $\frac{1}{4}$ inch of mercury in the syphon gauge. But one cannot fix the *quantity* of water to be frozen and the *time* to effect the freezing without knowing the power of the pump, the size of the receiver, the strength of the acid, the temperature of all parts of the apparatus, of all the materials to be used, and of the room in which the experiment is to be made. Such details cannot be entered into in this note.

Some readers may, perhaps, consider that the exhaustions quoted in the above table do not indicate very accurate pumps; for it is very frequently stated in books that double-barrelled pumps will exhaust to 1 in 1000, when the mercury in the gauge will be at $\frac{1}{100}$ inch. No doubt, air-pumps at three times the cost of any in this list could be made a little more accurate than these are: but I fancy that when the above statement is made as regarding ordinary working air-pumps, due care has not been taken to distinguish the power of the pump from the fallacy of the gauge.

RECEIVERS *of various forms and sizes suitable for experiments with the different air-pumps* are described between Nos. 800 and 900, and also in connection with the principal experiments.

The pumps are without receivers or other separate apparatus, at the quoted prices, unless otherwise described.

503. *Air-pump*, double barrels, 6½ inches long, 1½-inch bore, 4¾-inch stroke, with 8-inch plate, on mahogany stand, and stopcock between the plate and the barrels. Fig. 503. *Price* 8*l*. 8*s*.

Exhausting power of this pump, see § 502.

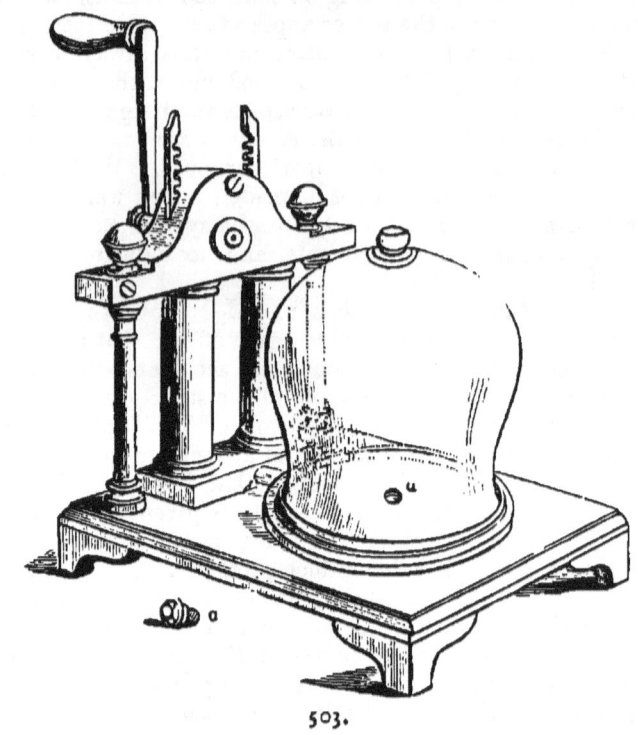

503.

504. *Air-pump*, double barrels, 7 inches long, 1¾-inch bore and 5½-inch stroke, with 10-inch ground plate, on mahogany stand, supported by four pillars, with small gauge plate, mercury gauge, and key. Fig. 504. *Price* 14*l*.

Exhausting power of this pump, see § 502.

505. *Tate's Double-action Air-pump*, which contains two pistons in one barrel, takes in the air from the receiver in the centre, and expels it at the two extremities of the barrel. Length of barrel, 16 inches; bore, 1¼ inch; stroke, 8½ inches. It has a 7-inch plate, a syphon-gauge, marked *r*, and a screw, marked *n*, to enable the pump to be used as a condensing-pump. It is mounted with a massive brass clamp, by which it can be securely fixed to any solid table. Fig. 505. *Price* 3*l*. 13*s*. 6*d*.

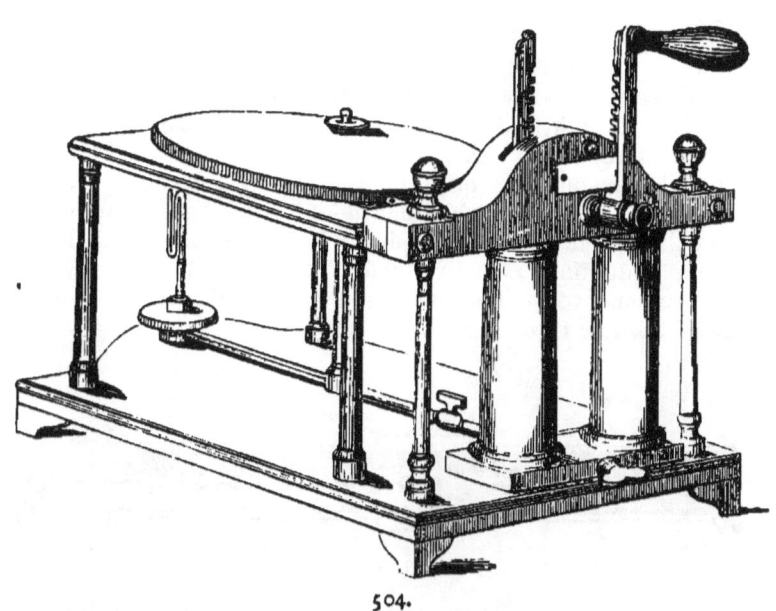

504.

A brass cap to collect the oil ejected at the valve *n*, is included in the price of the pump. It is shown at fig. 517.

Exhausting power of this pump, see § 502.

505.

It will freeze water over sulphuric acid in a receiver of 300 cubic inches, in 150 strokes at about 60° Fahr., and in half that number of strokes at about 40° Fahr.

506. This air-pump is made on the principle first explained by Mr. Tate (see his paper 'On a new double-acting air-pump, with a single cylinder,' in the 'Philosophical Magazine' for April, 1856). It is represented in perspective by fig. 505, and its barrel in section by figs. 507 and 508. The references in the following description are made to these three figures.

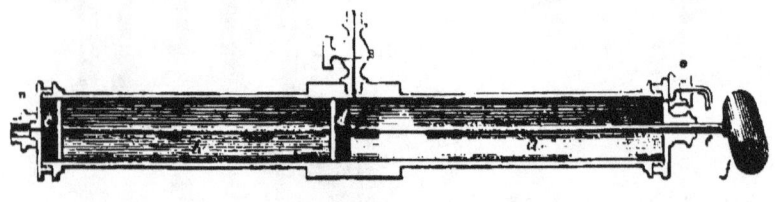

507.

The barrel a, b, is a brass cylinder, 16 inches long, with a bore of $1\frac{1}{4}$ inch diameter. It contains two pistons, c and d, figs. 507 and 508, which are both attached to one piston-rod, e, moved by the handle, f. The cylinder is firmly fixed in a horizontal position to a table by means of a massive brass clamp, g, fig. 505. Other methods of fixing it are shown in figs. 517 and 518. The pump-table, h, which is made 7 inches or more in diameter, is fixed above the middle of the barrel by the block i, and the stopcock k. There is an aircock at l, to let air enter into the receiver m, when required. At the end of the barrel n, there is a small orifice, with a valve opening outwards; this is covered by a brass cap, which has an external male screw. At the other end of the barrel, o, there is also a small orifice with a valve opening outwards, and covered with a brass cap terminating in a bent pipe p.

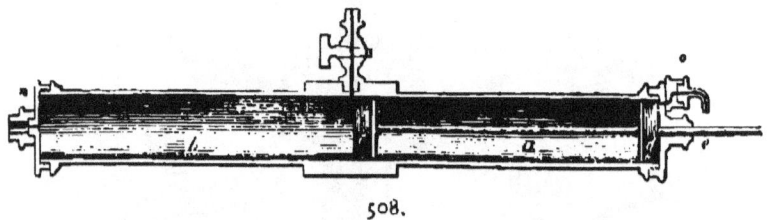

508.

509. *To ascertain if the Pump is in good working order.*—Clamp the pump firmly to a table, so that when the piston-rod is pushed in, and pulled out, by means of the handle f, as far as it will go, there is no vibration of the table h. If the pump works stiffly, pour a little neat's-foot oil into the hole in the middle of the table h, after removing

the syphon *r*, and opening the stopcock *k*. The oil will descend into the barrel and lubricate the pistons, and will afterwards be gradually forced out at both ends of the cylinder. The pipe *p* is so bent as to throw the ejected oil on the piston-rod, which must always be clean and well greased. The oil which is projected from the end *n* is collected in a brass box, which is represented in figs. 517 and 518, but not in fig. 505. This box must be often emptied.

It sometimes happens that when a pump is new the pistons *set*, or become *fixed*. In that case, after clamping the pump firmly to the table, you may pull out the piston-rod by main force. No harm will occur if you apply only as much force as is necessary for this purpose. The handle *f* must always be in a *horizontal* position, to be conveniently grasped by both hands; but as the piston easily turns round in the cylinder, the handle often comes into a vertical position. In that case, you must put it into the horizontal position by turning it *from left to right*, and not from right to left, otherwise you may unscrew the piston-rod from the piston.

Supposing the pump to be in working order, you must see that the plate *h* is clean, and that the receiver *m* is also clean, perfectly dry, and well greased on the flattened edge with tallow, or a mixture of wax and tallow, according to the temperature. This grease can be conveniently applied by means of the tallow-holder described at § 639.

When the receiver is thus placed on the table, it is easy to ascertain by a few strokes of the piston whether the junctions are all tight. If they are, the receiver soon becomes fixed to the table; if not, you must search for the leakage. All the parts of the pump should be tightly screwed up. Wherever there is a joint, there must be an intermediate oiled leather-washer, and this washer must be always clean, and soft with oil, and from time to time it must be examined, and if defective, renewed, § 611. When the leakage is not at one of the joints, it is commonly found to be between the receiver *m* and the table *h*. If it is owing to defective grinding of the edge of the receiver, and that defect is but slight, a little more tallow may cure it; but if the grinding has been insufficient, or if there is a chip in the glass, the rim must be re-ground.

If a few strokes of the piston are found to fix the receiver upon the plate, the pump is in working order. See § 711.

510. *Action of this Pump.*—The two pistons can be pushed into the positions shown by fig. 507. In this case there is a communication between the receiver *m*, and that half of the barrel which is marked *a*. I assume that the stopcock *k* is open, and *l* shut. This refers to fig. 505. The pistons can now be pulled into the positions shown by fig. 508. There is then a communication between the receiver *m* and that part of the barrel which is marked *b*. By this

second motion, all the air that was in the front half of the barrel *a*, is forced out of the barrel through the valve at the end, *o* —nearly all, but not quite all; for a small quantity of air remains in the little pipe between the piston *d* and the valve in *o*, and this expands into the half cylinder *a*, and into the receiver *m*, when the pistons are again put into the positions shown by fig. 507. In that third movement, the air contained in the half *b*, of the barrel, as shown in fig. 508, is forced out through the valve at the end *n*, except, as before, a small residue in the pipe between the piston *c*, and the valve in *n*. That this residue may be as little as possible, the pistons must at each movement *be driven quite home*. But this must be done firmly and steadily, not with too much violence or too much rapidity. The operator must remember that all the air that is expelled at each stroke has to pass through an opening which is, for the above reason, made as small as possible, and he must not give this little hole and the valve belonging to it too much work to do. If the barrel of the pump, and above all, the stuffing-box, *q* (fig. 505), becomes hot to the hand during the pumping, the operator is pumping too fast, and he must work more deliberately. Another precaution which he must take is, to work the piston-rod as evenly as he can in the direction of the cylinder, and not to make it waddle in the stuffing-box, *q*. If it waddles, the hole in the stuffing-box will speedily become enlarged, the pump will leak seriously, and the stuffing-box will require to be re-packed—work for the instrument-maker.

The exhausting power of the pump is tried by means of the syphon-gauge, which is marked *r* in fig. 505. That this may show the state of the exhaustion in the receiver *m*, a hole is bored in the brass foot by which the gauge is screwed into the hole in the table. With a receiver that is capable of holding 100 cubic inches of air, Tate's pump of the above size, and in perfect condition, will bring down the mercury to one-tenth of an inch in about 60 strokes. It will also readily freeze water over sulphuric acid in a flat receiver of 300 cubic inches; but the requisite number of strokes for this experiment varies greatly with the temperature of the water, and of the apparatus, and the apartment, from 150 strokes at between 60° and 70° Fahr., to half that number at between 30° and 40° Fahr.

The other pieces of apparatus shown in fig. 505 are, a screw, *s*, adapted to the hole in the table, *h*, and intended to prevent the running of water and mercury into the barrel of the pump, when spilt in certain experiments on the table, *h*; the jet *t*, and the water-pipe *u*, are for the experiment called *a fountain in vacuo*, which I shall describe presently.

511. *Use as a Condensing Pump.*—If a globular vessel mounted with a brass cap, containing a female screw, is adapted to the screw

at the end of the barrel *n*, the air which is there ejected from the pump will be necessarily forced into the vessel so placed to receive it.

512. *Advantages possessed by Tate's Air-pump.*—There is no valve placed between the receiver and the barrel of the pump, so that when the air becomes rarefied it is not required to lift a valve, but has simply to diffuse itself as each half of the barrel is alternately opened to receive it. The pump is not difficult to work. Though the barrel is 16 inches long, the effective stroke is only 8 inches. At first, the friction of two pistons makes the pull rather stiff, but as the exhaustion proceeds the pull becomes easier, because the action of the external atmosphere is cut off from the pistons by the valves placed at *o* and *n*, which is the reverse of what occurs with all pumps that have valves placed between the cylinder and the receiver.

Tate's Pump compared with Air-pumps with double barrels.— Though Tate's pump is a great improvement upon all single-barrel air-pumps, and does the work of exhaustion thoroughly, yet as its power is only in proportion to the capacity of its barrel, it does not work with sufficient expedition for a lecturer who desires to perform a series of experiments before a large, and perhaps an impatient, audience. When the size of the barrel is much enlarged, the labour is too heavy for the operator's hand. The piston-rod must be worked by a rack and pinion moved by a lever, and the apparatus then becomes expensive.

To meet the necessity of a more rapid, though less effectual exhaustion, recourse is had to air-pumps with double barrels, two common forms of which are represented in figs. 503 and 504.

Fig. 504 is a pump with a raised plate and a syphon-gauge. Fig. 503 is a pump of a cheaper construction without a syphon-gauge. It is also represented in the figure without a stopcock between the receiver and the barrels, which is a very unadvisable piece of economy, because it presupposes an absence of leakage at all joints of the pump, which, though desirable, is not always obtainable, especially when the pump is old. In fact, this pump is now supplied with a stopcock.

In pumps of this description, with vertical barrels, there are valves in the pistons, and also at the base of each cylinder, and the exhausting action of the pump ceases when the rarefied air in the receiver is no longer able to lift the lower valve when the piston is drawn upwards in the barrel. The labour of pumping increases with the exhaustion, and with wide barrels is considerable, because you have a vacuum under the piston and the full pressure of the atmosphere upon it, whereas in Tate's pump the pressure of the atmosphere is cut off from the pistons by the valves, which open outwards only, at the two ends of the barrel.

When rapid action and complete exhaustion are both required, it is advisable to use the large size of Tate's pump, No. 519.

515. *Freezing of Water.*—With the help of Tate's air-pump, the porcelain acid-pan, fig. *j*, and the flat receiver, fig. *k* 518, it is easy to freeze water. The pan must be half filled with concentrated sulphuric acid, not the fuming Nordhausen acid, but oil of vitriol that has not been diluted. The water should be put into a watch-glass, placed upon the acid-pan.

This experiment succeeds best when the pump, the water, and the air of the room are as cold as possible. The pump and every part of the apparatus being in good condition, it takes twice as many strokes of the piston to freeze water when the air is at 60° Fahr. as it takes when the air is at 40° Fahr. In winter, when the apparatus and water are tolerably cold, Tate's pump will freeze the water with less than 100 strokes. In summer, at about 60°, it will require at least 150 strokes; and if you allow the water and pump to stand in the sun till they are warm, or if you take diluted acid, or too much water, or too large a receiver, you will entirely fail to freeze the water.

A small quantity of water is, of course, more easily frozen than a large quantity; but when Tate's pump (the small size) is in good condition, and the weather cold, three or four ounces of water can be frozen, if placed in a thin porous earthenware capsule.

As Tate's pump is usually sold with adjuncts for producing a fountain in vacuo, I shall add a description of that experiment.

516. *Fountain in Vacuo.*—The table *h* of Tate's air-pump can be unscrewed in company with the stopcock *k*, from the block *i*, fig. 505. The jet *t* can be screwed into the hole in the middle of the plate *h*. The pipe *u*, fig. 505, or *a*, fig. 516, can be screwed to the stopcock *k*. These letters refer equally to figs. 505 and 516.

Process.—The pump being in good working order, ascertain that the table *h* and stopcock *k* are easily removable. Screw up tight, put in the jet *t*, cover with the conical glass receiver B; or, in default of that, with the cylinder of the guinea and feather glass closed at top with a well-greased glass

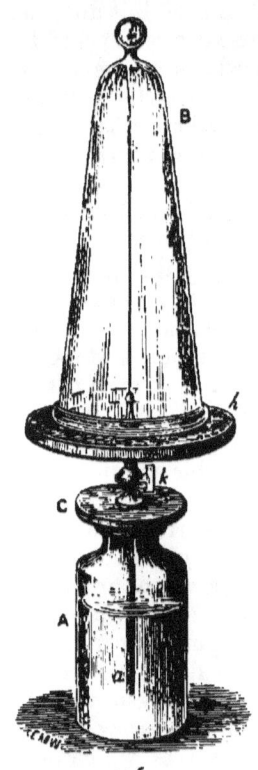

516.

plate. Have ready a wide-mouthed bottle or jar, A, filled with water nearly to the neck; also a thin disk of glass, wood, or metal, with a hole in the centre, c. Exhaust. Close the stopcock k, and then unscrew it, and all above it, from the pump. Put on the disk c, screw on the pipe a, and place the whole upon the glass bottle A. If the stopcock k is now opened, the weight of the atmosphere acting on the surface of the water in the vessel A forces the water, in the form of a jet or fountain, up into the exhausted receiver B.

As this experiment wets the plate and stopcock of the pump, and renders them unfit for other experiments until thoroughly dried, it is better to use for it a separate small table and stopcock, which is commonly called a transferer. See §§ 627 and 462. When a teacher has no separate transferer, it is better to defer experiments with water till the other experiments of the same day's lesson have been performed, because it takes some time to dry the transfer-plate and the stopcock sufficiently to enable other exhaustions to be made by the pump; for the presence of vapour diminishes its power. See § 462.

517. *Tate's Air-pump*, of the same form and dimensions as No. 505, but mounted on a solid and elegant japanned iron pedestal, represented by a, fig. 517. Price of the pump on pedestal support, with syphon gauge, 3*l*. 13*s*. 6*d*.

The *prices* of the extra pieces represented in fig. 517 will be found at the following numbers:—*b*, at No. 561; *c*, at No. 562.

517.

The pedestal requires to be screwed to the table where the pump is to be used. The pump is then perfectly solid. On the other hand, the clamp (fig. 505) is useful when the pump has to be carried about for use in different localities.

The round box represented at the far end of the pump cylinder in figs. 517, 518, and 520, is intended to catch the oil, of which more or less is expelled from the pump with the air at every stroke. There is a hole at the upper part of the box to let out the air. From time to time the oil should be removed from the box. Though not shown at fig. 505, this oil-box is now sold with the pump, No. 505.

Exhausting power of this pump, the same as that of 505.

518. *Tate's Air-pump*, same form and size as No. 505, but mounted on a solid iron table, a, fig. 518, which has four legs screwed to an iron plate, b, which plate can either be fixed permanently to the work-table by four screws, or be fastened to it by the large iron clamp, c, which permits the removal of the pump. *Price* of the pump and table, a, b, with the common syphon gauge, d, fig. 517, 3l. 13s. 6d.

518 A. Extra fittings for any of Tate's pumps, as represented by c, d, e, f, g, g, h, i, j, k, l, fig. 518, cost 2l. 2s.

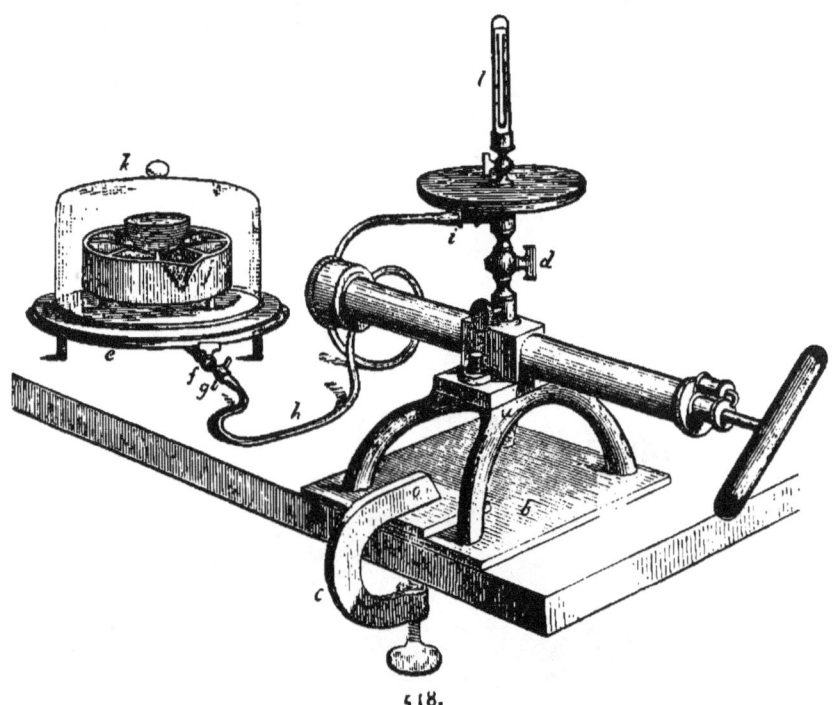

518.

Exhausting power of this pump, the same as that of Nos. 505 and 517. The three pumps are of the same size and power, they differ only in the style of mounting, and are supplied at the same price.

519. *Tate's Air-pump*, of the same form as the preceding, No. 518, but of above double the size and power. namely, with a barrel of 17 inches in length, 1¾-inch bore, and 5¼-inch stroke. It is mounted on an iron table similar to *a, b*, fig. 518, with a plate of 10 inches diameter, and supplied with the extra joint and arm marked *b, c*, in fig. 517. and the gauge marked *l*, in fig. 518. This is the largest and most powerful form of Tate's air-pump which can be conveniently worked without rack-work, lever, or other machinery. It gives great power, is pretty easily worked, and is sold at a moderate price, 8*l*.

Exhausting power of this pump, see § 522.

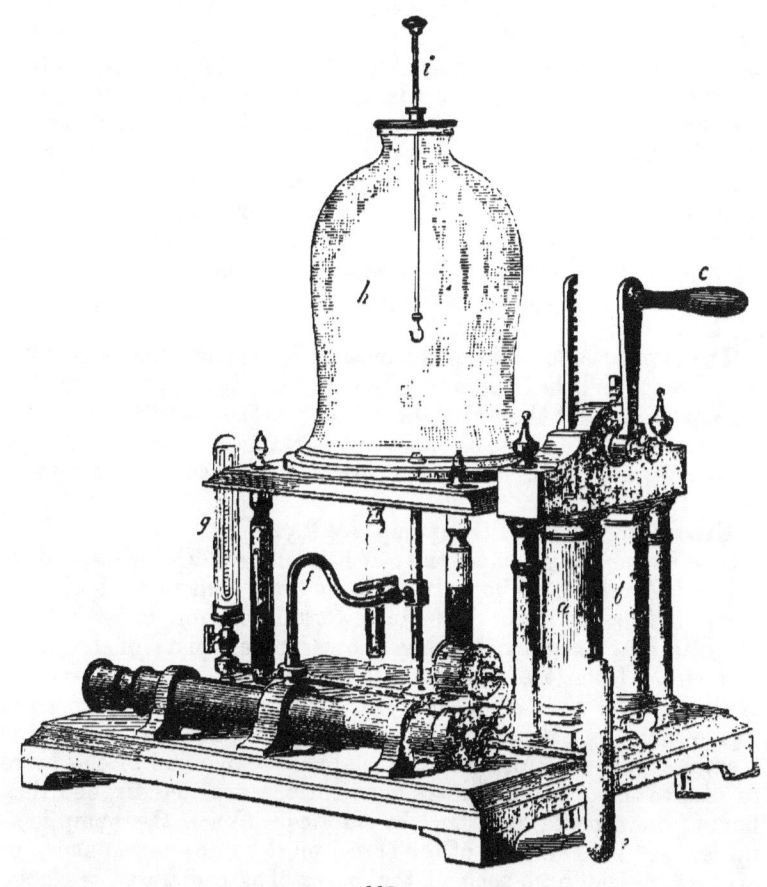
520.

520. *Air-pump with three Barrels*, serving either for rapid action at lectures, or for more complete exhaustion for researches, fig. 520. Price, without the jar and rod, h, i, 18l.

Exhausting power of this pump, see § 502.

This apparatus consists of an air-pump with two vertical barrels, marked a, b, in fig. 520, which are worked as usual by the handle c acting on rack-work. Bore of the vertical barrels, $1\frac{3}{4}$ inch; length of barrels, 7 inches; stroke, $5\frac{1}{2}$ inches; diameter of the pump-plate, 10 inches. With this arrangement, large receivers, having, for example, a capacity of a thousand cubic inches, can be rapidly and easily exhausted till the mercury in the gauge falls to $\frac{3}{4}$ inch, and small receivers can be brought to a vacuum of $\frac{1}{4}$ inch. This is sufficient exhausting power for most of the experiments that are usually exhibited at lectures to illustrate the principles of Pneumatics. But, as more perfect exhaustion is sometimes required, a separate pump on Tate's plan is added to the vertical pump. When the power of the latter ceases, Tate's pump is put into action, and the exhaustion then proceeds, until the mercury in the gauge descends in receivers of a thousand cubic inches to $\frac{3}{16}$ inch, and in small receivers to $\frac{1}{8}$ inch, and even to $\frac{1}{20}$ inch. See the experiments recorded at § 502. This compound air-pump is therefore adapted either to give quick results when moderate exhaustion is requisite, as at lectures, or more effectual exhaustion when that is required for special experiments or for researches. The extent of the exhaustion is shown by the gauge g, on the plan of No. 577, fixed on the lower table of the pump.

The apparatus is made in the most solid manner, and is mounted on a French-polished mahogany frame. The price quoted includes the gauge, but not the rod and receiver marked h and i.

521. *Tate's Air-pump*, of large size, arranged for easy and rapid action, worked by a winch and crank, regulated by a fly-wheel, fig. 521. Price 21l.

Exhausting power of this pump, see § 502.

The barrel is 12 inches long; it has a bore of $2\frac{1}{2}$ inches, and the stroke is 6 inches. The valves are of brass, and work in oil; the barrel is fixed upright. The framework is of iron, and the top is of polished mahogany, bearing a brass pump plate of 10 inches diameter. Upon the mahogany top there is a screw to receive a syphon-gauge, similar to No. 577. There is also a descending gauge-tube of 32 inches long below the pump, and accompanied by a scale of inches. An oil-box is adapted to each end of the barrel to receive the oil which passes out with the expelled air, and these boxes from time to time must be emptied. When the pump is set up for use, the two ends of the barrel must be unscrewed and some oil must be put into each of the boxes that contain valves, about

AIR-PUMPS. 125

4 fluid ounces in the lower box, and 2 fluid ounces in the **upper box**. About 2 fluid ounces of oil should also be put into the pump by the hole in the centre of the ground plate, which, on working the handle of the pump, will be distributed throughout the interior.

521.

An air-pump of this description, which effects with ease in five minutes, in large vessels, a vacuum represented by 1/4 inch of mercury in the gauge, can be usefully applied in many of the arts.

522. *Single-barrel Air-pump*, the barrel of brass, mounted on a mahogany stand, form of fig. 522, but without the basin and receiver. Diameter of the plate, 6¾ inches; length of the barrel, 8 inches; bore of the barrel, 1¼ inch; length of stroke, 7 inches. *Price, with stopcock,* 45s.

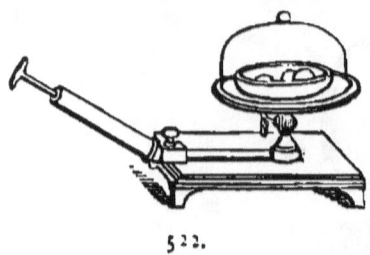

522.

523. *Small Single-barrel Air-pump*, the barrel of brass, mounted on a mahogany stand, form of fig. 522, but without the basin, the receiver, and the stopcock. Diameter of the plate, 4¼ inches; length of the barrel, 6 inches; bore of the barrel, 1 inch; length of stroke, 6 inches. *Price* 21s.

527. *Remarks on the comparative labour that attends the working of these Air-pumps.*—I have stated at § 502, the comparative *exhausting-power* of these six air-pumps; and I will now say a few words on the comparative *labour* attending their use.

The air-pumps of the old description, with vertical barrels, such as Nos. 503, 504, and 520, are easy to work *at the beginning* of each exhaustion, and gradually become more difficult as the exhaustion approaches completion. This arises from the increasing pressure of the atmosphere upon the upper faces of the pistons in the barrels when the spaces under the pistons are nearly free from air. With Tate's pump the contrary effect is produced: at the beginning the pressure is considerable, but after two or three strokes it gradually becomes easier and easier, the atmosphere being cut off from the pistons by the valves at the ends of the barrel. Pistons of 1¼ inch diameter, such as belong to the pump No. 505, are easy to work even at the beginning of a process. But the pump No. 519, which has pistons of 1¾ inch diameter, requires for the first two or three pulls a pretty strong hand; for, besides the atmospheric pressure, there is in this case a much greater amount of friction from the pistons to be overcome: after two or three pulls, it works easily enough. With a view to apply Tate's principle to large pumps, the apparatus No. 521 has been made. In this case the force necessary to work the pistons is acquired mechanically, and we have an apparatus that is worked with great facility. It will be observed in the table given at § 502, that there are three pumps, Nos. 519, 520, 521, having 10-inch plates, which can be used with receivers that hold 1000 cubic inches of air, and which exhaust these cylinders with the same degree of accuracy, while the respective cost of these pumps is 8*l*., 18*l*., and 21*l*. The higher prices include the cost of the machinery for making the pumps work easily and quickly. The cheapest of these pumps will do the same work as the dearest, if you give it a little more *time* and a little more *labour*. The choice of a pump must be guided by circumstances. If you require rapid exhaustion with easy work, No. 521 is the best pump; and next to it, No. 520; but if economy is an object, the pump No. 519 may be safely taken. Comparing together two pumps of the same price—namely, Nos. 503 and 519—the former is easier and quicker in action; the latter is much more effective, and without any other drawback than the two or three stiff pulls at the beginning of an exhaustion. The small Tate's pumps, Nos. 505, 517, 518, are beyond question greatly superior to all the two-barrelled pumps that are usually sold at twice their price. These small pumps can be worked without difficulty, and, as shown in the table at § 502, they give excellent results.

CARE OF AN AIR-PUMP.

528. It should be made perfectly clean before being put away after having been used for a series of experiments. If it cannot be preserved in a glass case, a cover of pasteboard should be prepared to shelter it from dust. In winter it must not be taken out of a cold room and immediately submitted to use. It ought to be put for some time in a warm room. In pumping, the action should be moderate and regular. Every stroke should go home, so as to leave no residue of condensed air in the barrel; but the pumping must never be so rapid as to leave too little time for the condensed air to escape through the valves; it must never make the barrel of the pump warm; it must never make the mercury in the gauge jump about violently. Air, when let into a receiver by the aircock, must be let in very slowly, otherwise the action risks the spoiling of the gauge—the long barometer gauge is especially liable to destruction by a careless admission of air. A receiver should never be removed from a pump-plate without first having air let in by the aircock to set it free, and when the receiver is removed it should never be by a direct upward *pull*, but by a sliding motion towards the edge of the pump-plate. When a pump is set aside, all the stopcocks should be left open; for they are less subject to corrosion from acid vapours when open than they are when closed.

AIR SYRINGES, POLISHED BRASS.

Exhausting Syringes, without stopcocks, fig. 531 :—
531. Barrel 6 inches long, ¾ inch outside diameter. *Price* 8*s*.

532. Barrel 8 inches long, 1¼ inch diameter. *Price* 16s.
533. Barrel 9 inches long, 1¾ inch diameter, mounted on a clamp. *Price* 35s.
534. Barrel 12 inches long, 2 inches diameter, mounted on a flange to screw to a table. *Price* 42s.
535. Barrel 12 inches long, 2¼ inches diameter, mounted on a flange. *Price* 50s.

Condensing Syringes, without stopcocks, fig. 531 :—

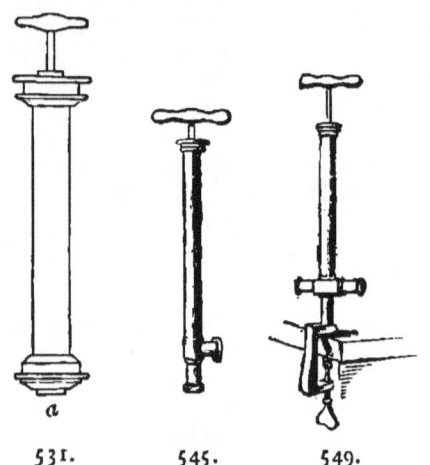

531.   545.   549.

541. Barrel 6 inches long, ¾ inch diameter. *Price* 8s.
542. Barrel 8 inches long, 1⅓ inch diameter. *Price* 16s.

Exhausting and Condensing Syringes, form of fig. 545 :—

545. Barrel 6 inches long, ⅞ inch diameter. *Price* 10s. 6d.
546. Barrel 8 inches long, 1⅓ inch diameter. *Price* 18s.

Exhausting and Condensing Syringe, with a cross-piece and clamp to fasten it to a table. fig. 549 :—

549. Barrel 6 inches long, ⅞ inch diameter. *Price* 18s.
550. Barrel 8 inches long, 1⅓ inch diameter. *Price* 27s.

All these syringes end with female screws, and therefore require stopcocks and connectors with male screws, to connect them with other apparatus. The woodcuts from No. 590 *to* 622 *will explain this matter fully.*

Extra Fittings for Tate's Pump.

560. *Extra Screw* between the pump-plate and the stopcock, represented at the upper part of *b* in fig. 517, and *d* in fig. 518, with a blank nut to close it when not required for use. *Price* 3s. 6d.

This is not a separate piece of metal, but a prolongation of the stopcock marked к in fig. 505. It is useful for the purposes represented in figs. 517 and 518.

561. *Extra joint with Screw.*—Those who already possess Tate's pump, as figured at No. 505, can have an extra joint with this screw supplied at the cost of 5s.

562. *Arm to carry a Syphon Gauge,* applicable to the extra screw No. 560, represented by *c*, fig. 517. *Price* 4s.

EXTRA FITTINGS FOR AIR-PUMPS.

The syphon-gauge, when used merely to test the power of a pump, can be screwed into the plate, as shown at r, fig. 505. In that case a hole is drilled in its brass base to permit the passage of the air. But when it is desired, during the progress of an experiment, to know the extent of exhaustion within the receiver, the gauge can be mounted, as represented by fig. 517. The gauge d must then be without the extra hole in its brass base. Instead of this form of gauge, that represented by fig. 577 can be used in this manner.

563. *Extra Pump Plate*, for use in drying chemicals in vacuo, or in freezing water, over sulphuric acid, consisting of a cast-iron table, mounted on three legs, with plate-glass surface, an air-tube and stopcock, the thread of which fits a union joint, as represented by e, f, g, fig. 518, three sizes:
563. 8-inch plate, with stopcock. *Price* 14s.
564. 10-inch plate, with stopcock. *Price* 18s.
565. 12-inch plate, with stopcock. *Price* 25s.
566. *Connecting Tube*, flexible metal, 3 feet long, h, fig. 518, with a screw i, to fit the extra joint d, and a union joint g, to fit the stopcock f, of the extra plate. *Price* 4s. 6d.

One connecting-tube serves for any number of separate plates, Nos. 563 to 565, each of the plates having a stopcock to keep the vacuum.

During the exhaustion of a receiver placed over a separate pump plate, a gauge is fixed in the centre of the fixed pump plate, as represented by fig. 518, to mark the progress of the exhaustion.

567. The connecting-tube, with the stopcock f, g, h, i, fig. 518, complete. *Price* 8s.

568. Flat glass receivers for the extra pump plates, of strong German and Bohemian glass, welded and ground on the edges, see article on Receivers, § 796, and fig. 800.

569. Pans for containing sulphuric acid, represented at fig. 518, are described at § 705.

Syphon Gauges, for showing the extent of exhaustion effected by air-pumps within their receivers. These are of three kinds:—

575. Common three-limbed syphon gauge, mounted on a brass foot, with male screw, fig. 575. *Price* 3s. 6d.
This gauge may be had either with or without an air-hole drilled in the brass foot; see Note to No. 562. It is used as shown by r, fig. 505, and d, fig. 517.

576. Syphon gauge, with scale, form of fig. 576, mounted on a flat stand, and requiring to be placed under a receiver for trial. *Price* 5s.

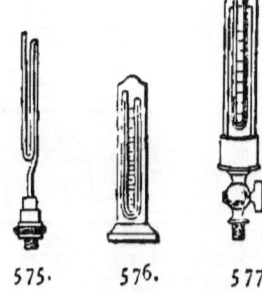

575.   576.   577.

577. Syphon gauge, with glass scale, form of fig. 577, the gauge enclosed in a glass tube. *Price*, with stopcock, 10s. 6d.

578. Ditto, without stopcock. *Price* 7s.

Of these three syphon gauges, the one that is least trustworthy is the first, No. 575, in consequence of the difficulty of filling it with mercury quite free from air. This variety also is not graduated. The difference in the level of the two columns of mercury indicates the extent of the exhaustion. The other two are graduated to show twentieths of an inch. No. 577 is represented in use by l, fig. 518, and g, fig. 520.

Purchasers of the pumps, Nos. 505, 517, 518, with which the gauge No. 575 is delivered, may have No. 577 instead, on payment of the difference in price.

The precautions to be taken to secure accurate results in the use of syphon gauges have been detailed at § 502.

STOPCOCKS, CONNECTORS, AND FITTINGS FOR OCCASIONAL USE.

The following stopcocks and connectors are all London-made, of the best quality, and of polished brass:—

Stopcocks and connectors of polished iron cost about *one-half more* than those of polished brass.

Stopcocks.

590. Stopcock, with two male screws, fig. 590. *Price* 3s.

591. Stopcock, with one male and one female screw, fig. 591. *Price* 3s.

590.   591.

592. Stopcock, with a male screw at one end, and at the other end a union joint and a long brass tube for attaching a flexible tube, fig. 592. *Price* 5s.

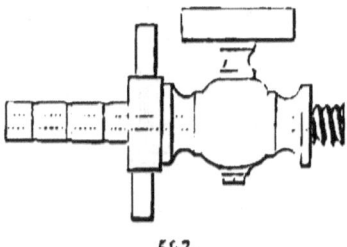

592.

594. Stopcock, with a female screw at one end, and at the other end a union joint and a long brass tube for attaching a flexible tube, similar to fig. 592, but having a female screw instead of a male screw. *Price* 5s.

EXTRA FITTINGS FOR AIR-PUMPS.   131

595. Stopcock, with a tube at one end for connecting a caoutchouc tube, and at the other a male screw, fig. 595.   *Price* 3*s*.

596. A similar stopcock and tube, but having a female screw at one end.   *Price* 3*s*.

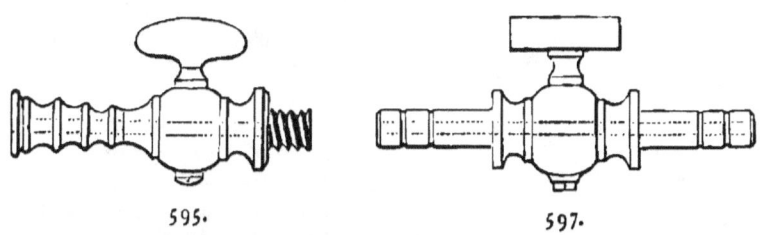

595.   597.

597. Stopcock, having at each end a tube for connecting a caoutchouc tube, fig. 597.   *Price* 3*s*.

Connectors.

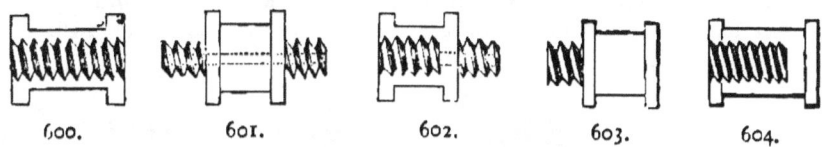

600.   601.   602.   603.   604.

600. Connector, with two female screws, fig. 600.   *Price* 1*s*.
601. Connector, with two male screws, fig. 601.   *Price* 1*s*.
602. Connector, with one male and one female screw, fig. 602.   *Price* 1*s*.
603. Blank nut, with one male screw, fig. 603.   *Price* 1*s*.
604. Blank nut, with one female screw, fig. 604.   *Price* 1*s*.

These blank nuts are used to stop openings that are not required; see No. 560.

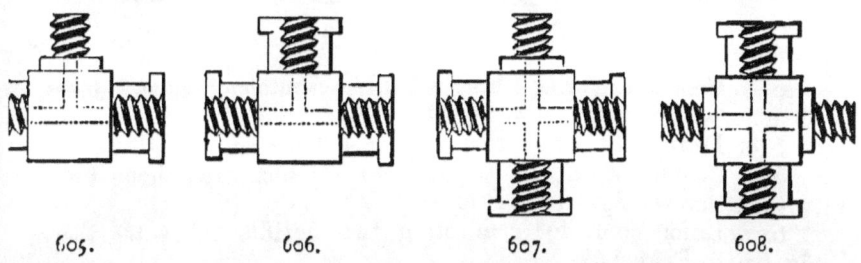

605.   606.   607.   608.

Three-way and four-way Connectors; four patterns. *Price each* 3*s*. :—

605. Connector, two female screws, one male screw, fig. 605.

K 2

606. Connector, three female screws, fig. 606.
607. Connector, three female screws and one male screw, fig. 607.
608. Connector, two female screws and two male screws, fig. 608.
609. *Brass Caps* for bell-jar receivers, and for globes for gases, with female screw; diameters to suit glass necks of ¾, 1, 1¼ inch, fig. 609. *Price each* 1s.

610. Similar brass caps, 1½, 2, and 2½ inches in diameter, to suit large receivers. *Price each* 2s.

611. *Washers.*—When stopcocks, connectors, &c., are screwed together, there should be a piece of oiled leather, a *washer*, of the form of fig. 611, placed between them, to ensure an air-tight joint. These washers must always be kept clean and softened with oil, and the screws should be frequently cleaned and slightly oiled. *Price per dozen* 9d.

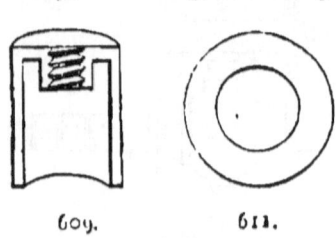

609.   611.

616. Bladder piece, or socket to tie in the neck of a bladder or a gas bag, with female screw for receiving a stopcock, fig. 616. Price 9d.

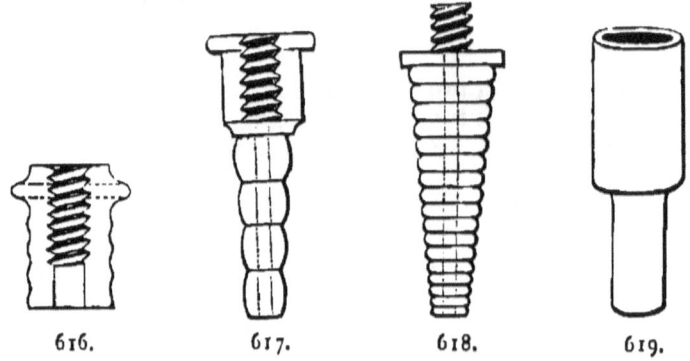

616.   617.   618.   619.

617. Connector for attaching a flexible caoutchouc tube to brass fittings, with female screw, fig. 617. *Price* 9d.

618. Ditto, with male screw, fig. 618. *Price* 9d.

619. Connector to join a ¼-inch to a ½-inch caoutchouc tube, without screws, fig. 619. *Price* 6d.

620. Union joint for connecting two flexible tubes together, fig. 620. *Price* 2s.

621. Block to be screwed to a table, with female screw to receive the ends of syringes, cross-pieces, or other articles that need to be fixed in an upright position; two kinds, fig. 621, flat. *Price* 1s.

622. Ditto, raised. *Price* 1s.

EXTRA FITTINGS FOR AIR-PUMPS. 133

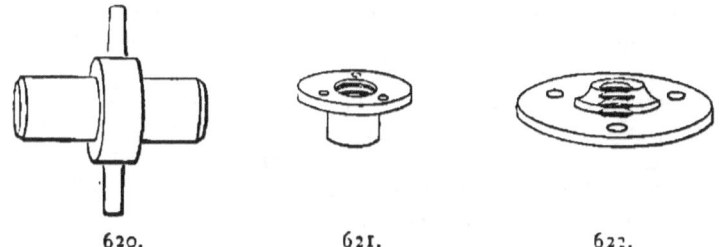

620.   621.   622.

624. *Ground Brass Plate and Hook* for suspending objects in exhausted receivers that have welted and ground necks, such as Nos. 684, 725, 4 inches diameter. *Price* 5s.

625. *Ground Brass Plate, with sliding Rod,* passing through a stuffing-box, to afford the means of moving apparatus suspended in an exhausted receiver; see Nos. 696, 666. 4-inch plate. *Price* 12s.

The brass plate has a small hook on the under side, which is not shown in fig. 625. The hook at the bottom can be unscrewed and removed, for certain operations.

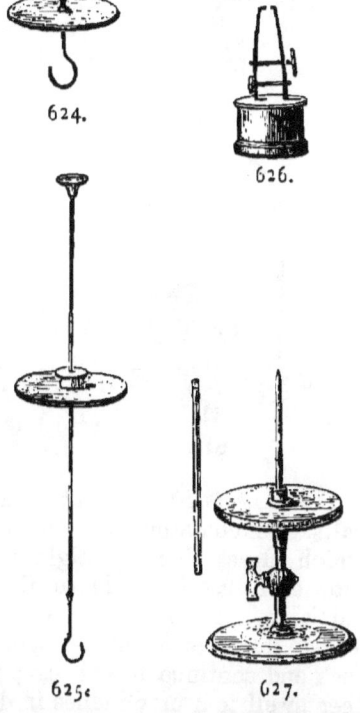

626. *Clip and Weight* for holding plants under water in vacuo, to show the quantity of air disengaged, fig. 626. *Price* 2s. 6d.

The clip in action is shown by fig. 676.

626 A. More powerful clip for large objects, with box containing lead shot. *Price* 7s. 6d.

627. *Single Transfer Plate, or Fountain Plate;* a ground brass plate, with stopcock, water pipe, and fountain jet, for producing a fountain in vacuo, as described at § 462. Fig. 627 shows the apparatus mounted on a mahogany stand; the small figure is the water-pipe.

627. A. 4-inch plate, *price* 15s. B. 5-inch plate, *price* 18s.

628. *Experiment with the single Transfer Plate.*—The transfer plate is used when it is necessary to produce a vacuum in a vessel suddenly. The arrangement is represented by fig. 628. The

small upper receiver is the vessel in which the vacuum is to be produced suddenly. Under it is a larger receiver, and the transferer connects them. The subject to be operated upon is placed in the upper receiver, the lower receiver is then exhausted; after which, on turning the stopcock of the transferer, the air rushes suddenly from the upper into the lower receiver, leaving a partial vacuum.

The degree of rarefaction will be in proportion to the difference in the capacities of the two receivers.

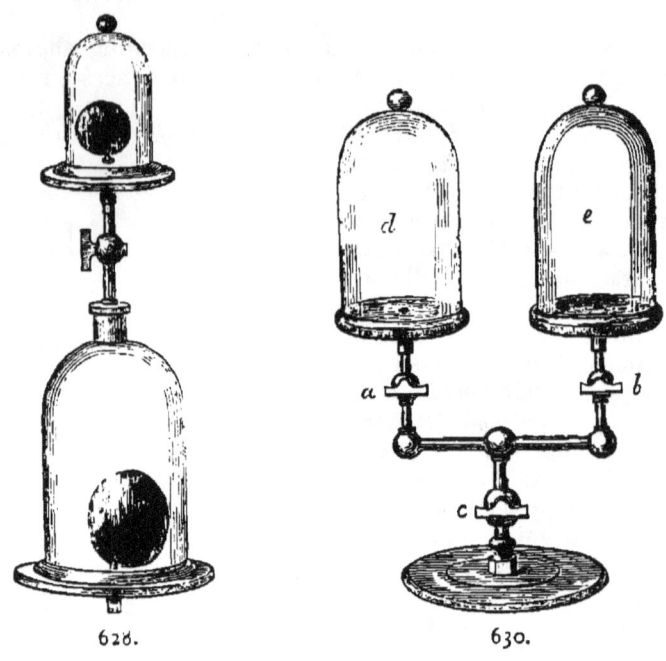

628.   630.

Experiment with Caoutchouc Balls.—Take two small caoutchouc balls, each containing a little air. Dip the mouths into chloroform, which closes them air-tight without tying them. Put one ball into each receiver; leave the transfer stopcock open; exhaust partially. As the air passes out, both caoutchouc balls swell. When they are about $1\frac{1}{2}$ inch in diameter, shut off the transfer cock and continue to exhaust; the ball in the lower receiver will then swell to 4 or 5 inches in diameter. Cease to pump; open the transfer cock, upon which the upper ball will suddenly become larger and the lower ball smaller, because the air rushes from the upper receiver to act in the lower receiver.

Price of the receiver, mounted with screw cap, 5s.

630. *Double Transferer,* fig. 630, consisting of two 4-inch ground

brass plates, each with a stopcock below, both connected with a horizontal tube, which is attached in the centre to a third stopcock, capable of being screwed to a stand, as here figured, or to the table of the air-pump. Hence receivers, such as *d*, *e*, placed upon the transfer plates, can be exhausted separately, or be put into communication with one another. *Price of the set, without receivers,* 40s.

Unscrew the apparatus from the wooden foot, and screw the end of the stopcock *c* into the plate of the air-pump. Open the stopcocks *a* and *c*, and close the stopcock *b*. Place a receiver *d* upon the table of *a*, exhaust the air from it, and turn off the stopcocks *a* and *c*. The apparatus can then be removed from the pump, and the wooden foot be screwed to it. Then put a receiver *e* upon the plate over the stopcock *b*. The apparatus will now appear as represented in fig. 630, but the receiver *d* will be fixed, and the receiver *e* be loose. Next, open both *a* and *b*. Half the air contained in the receiver *e* will then pass into the receiver *d*, and both receivers will be fixed upon their plates by the atmospheric pressure, though each will be kept down by only one-half of the pressure that acted on the first receiver when exhausted of its air. This experiment shows very clearly the expansibility of air.

631. *Aurora Borealis Apparatus*, consisting of a ground brass plate, 4 inches diameter, with three spikes; the plate adapted in width to the top of the conical glass receiver of the guinea and feather apparatus; and of a second brass plate, with three spikes, to screw into the plate of the air-pump. *Price of the pair* 10s. 6d.

When this apparatus is properly arranged, screwed up tight, and exhausted, the aurora borealis appearance is produced when a current of electricity is sent through the vacuum.

634. Brass clamp, for fastening an air-pump to a table, &c., 2½ inch, fig. 634. *Price* 3s. 6d.

635. Iron clamp, fig. 634, 3¼ inch. *Price* 1s. 6d.

636. Iron clamp, 4¼ inch, represented by fig. 636, and by *c*, fig. 518. *Price* 4s.

637. Brass key, for screwing up the joints of air-pump apparatus, connectors to stopcocks, &c.

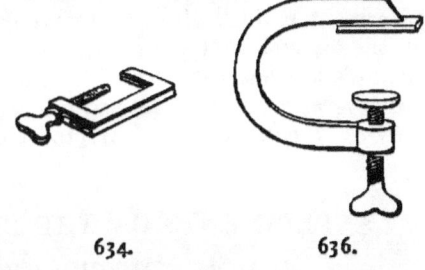

634.   636.

A. Single, *price* 1s. 6d.   B. Double, *price* 2s. 6d.

Tallow-holder, which consists of a mahogany tube, containing a piston moved by a screw, fig. 638.

638. Small tallow-holder for general use, bore ⅝ inch. *Price* 1s.

639. Large tallow-holder, for use with the air-pump, bore ⅜ inch. *Price* 1s. 6d.

Tallow, or a mixture of tallow and wax in summer, or tallow and lard in winter, is convenient for greasing the edge of a glass vessel previously to decanting a liquid, in order to prevent the running of the liquid over the edge of the vessel so as to descend outside.

638.

Tallow is also required to grease the edge of air-pump receivers, to make them fit the ground-plate air-tight. To fill the tallow-holder, the piston is screwed back to the top of the tube, and melted tallow is poured in till the tube is full. After cooling, the tallow is projected as required by turning the screw.

640. Porcelain pourer, in which to mix and melt tallow, wax, &c., to fill the tallow-holder. *Price* 1s.

640.

In collecting tallow from the air-pump plate and the rims of receivers, which can be done by a steel spatula No. 640 A, and transferred to the pourer No. 640, care must be taken to collect no dirt or grit, which would subsequently act mischievously. After melting such collected tallow, it ought to be examined for grit by the finger before it is applied to use. Fat set aside for future use should be preserved in a pot provided with a cover to keep it free from dust.

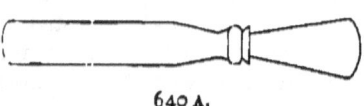

640 A.

641. *Grooved Plate.*—In the middle of every air-pump plate is a hole through which the air is drawn in exhaustion. Many objects with flat bottoms are placed upon this plate, and may stop up the hole and prevent exhaustion, unless that stoppage is guarded against. Sometimes a hole is drilled in the flat foot of the object subjected to experiment, or in the tube that rises from it, to retain a thoroughfare for the air undergoing exhaustion. A very useful foot for general purposes is a flat plate of metal, with parallel faces, and on the under face several grooves leading to the central hole in the pump plate.

Price of 3-*inch or* 4-*inch grooved plates,* 4d.

EXPERIMENTS ON THE PROPERTIES OF AIR.

IN FOUR GROUPS.

GROUP A.—THE WEIGHT AND RESISTANCE OF AIR, § 650 TO § 670.

650. *Estimation of the Specific Gravities of Gases.*—The apparatus employed for this purpose is represented by figs. 650 and 651.

EXPERIMENTS ON THE WEIGHT OF AIR. 137

650. Tall glass gas receiver, about 12 inches high by 6 or 7 inches wide, flanged and ground at the bottom, graduated into (about) 350 cubic inches, mounted with a stout glass globe, 5 inches diameter, two stopcocks with double male screws, and a double female connector. *Price of the set*, fig. 650, 20s.

651. Flask to show that air has weight, namely, a globe of very light glass, mounted with brass cap, with a small stopcock, an extra screw, a, to adapt it to the air-pump, and a hook for the balance. Size of the globe, about $4\frac{1}{2}$ inches diameter; capacity, about 50 cubic inches; for use in weighing air and gases. *Price* 7s. 6d.

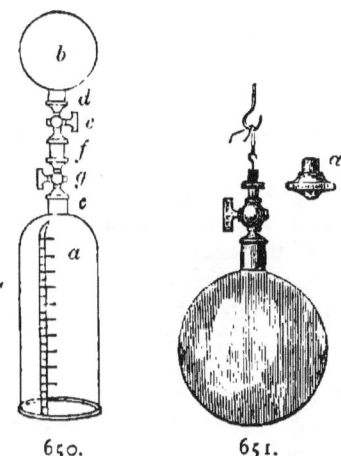

650. 651.

In fig. 651, the screw marked a serves either to attach the globe to the air-pump or to the jar 650 a.

I copy the following details from GRIFFIN's 'Chemical Recreations,' page 277. The apparatus for estimating the specific gravities of gases consists of the following parts:—

" 1. A cylindrical jar, a, in which the gas can be collected. It may be graduated into cubic inches and decimal parts, or into cubic centimetres.

" 2. A very light glass globe, b, fitted with a small and light brass cap, d, and a small stopcock e. In the figure, the caps d and c, and the stopcocks e and g, are made of the same size. The figure, however, represents an apparatus suitable for other experiments in chemistry. But for the determination of the specific gravity of gases, not only must the globe b be made as light as possible, but the brass fittings d and e must also be made small, and very light.

" 3. The brass fittings c to g complete this apparatus; c is a cap cemented to the jar a. It has a female screw; d, a cap cemented to the globe b, has also a female screw; f is a connector with two female screws; and the two stopcocks e and f have each two male screws. These pieces are shown by figs. 590, 600, and 609, at about half the full size; excepting that the small cap d should not be larger than fig. 609, and the stopcock e not larger than fig. 590.

" The operation of taking the specific gravity of a gas is performed as follows:—

" The globe, b, having been made perfectly clean and dry, is filled with perfectly dry atmospheric air, and with the cap, d, and stopcock, e, is weighed. It is then screwed to an air-pump, or air-syringe, and exhausted of the atmospheric air as completely as possible. The stopcock, e, is closed, and the globe is weighed again. The difference between the first weighing and the second shows the weight of the atmospheric air which has been withdrawn by the air-pump. The exhausted globe, b, is next connected to the receiver, a, by the intermediate brass-work shown in the figure. The two stopcocks, e, g, being then opened, gas passes from the receiver, a, into the globe, b, and fills it. The stopcocks are then closed, the globe, with its stopcock, e, is unscrewed from f, and once more weighed. The difference between the result of this weighing and of the second weighing shows the weight of the gas submitted to trial.

"If the jar a, from which the globe is filled with gas, stands over water, the gas will be saturated with aqueous vapour, the quantity of which must be allowed for by calculation, or the gas must be passed, for weighing, not directly from the jar a, into the globe b, but through an intermediate apparatus for drying it.

"If the globe b, employed for these experiments, be sufficiently large to contain, at 60° Fahr., and 30 inches bar., 46·7 cubic inches of gas, the bulk will represent one grain of hydrogen gas, and that globe will contain the quantities of the elementary gases which are represented by the atomic weights of these gases. That is to say—

"1 volume of hydrogen gas being = 1 grain
1 volume of oxygen gas will be = 16 grains
1 volume of nitrogen gas ,, = 14 ,,
1 volume of chlorine gas ,, = 35·5 ,,

And the quantity of a compound gas will be represented by its atomic weight divided by its atomic measure. Thus:—

"1 volume of carbonic acid gas, CO^2, will be = 44÷2 = 22 grains.
1 volume of carbonic oxide gas, CO, will be = 28÷2 = 14 grains.

The same volume of atmospheric air will be 14·47 grains."

The light glass globe usually sold with the apparatus No. 651, commonly contains about 50 cubic inches, or 15 grains of air, so that the large balance No. 122 can be used to weight the globe No. 651, and give a result approximately true.

652. *Experimental proof that Air has weight.*—The apparatus and experiment described at §§ 650, 651, afford the means of proving this fact in the most precise manner. The following experiment also serves to prove it, in an easy and satisfactory manner, when no air-pump is at command. Connect a narrow glass tube, by means of a sound cork, to a globular flask, draw out the external end of the tube to a very fine point, which must be left open. The joints

must be completely air-tight, and the cork may be varnished to ensure this condition. The apparatus being prepared, the flask is to be gradually heated over a spirit lamp or charcoal fire, and is finally to be made very hot, to drive out as much as possible of the included air. The point of the tube is then to be suddenly brought into the flame of a spirit lamp or a blowpipe jet, and closed by fusion. The best way to do this is to use a second spirit lamp and blowpipe, while the apparatus is being heated over a separate lamp. Allow the apparatus to cool, and counterpoise it in the pan of a balance. Then break off the point of the tube, upon which air will rush into the flask, the apparatus will become heavier and the counterpoise will be overbalanced.

652.

653. *Comparative Specific Gravities of Hydrogen Gas and Carbonic Acid Gas.*—These have been stated at No. 651, and the facts can be proved by collecting the gases in the cylinder No. 650, transferring them to the globe No. 651, and weighing them. But the following method is quicker and therefore more suitable for a class experiment. It is a rough sort of operation, but it demonstrates the prime facts that hydrogen gas is much lighter than air, and carbonic acid gas much heavier. Fig. 653 is a cylinder of pasteboard, measuring about 7 inches in length and 5 inches in diameter; it is open at one end and provided with strings, by which it can be suspended by either end to the hooked pan of the hydrostatic balance No. 122.

653.

Experiment A.—Suspend the cylinder *mouth downwards* and counterbalance it; then bring under it a jar or bottle filled with hydrogen gas. Remove the stopper from the gas bottle and pour the gas into the paper cylinder. In consequence of the lightness of the hydrogen gas, it will rise into the cylinder and the atmospheric air will sink down. The counterpoise will be thereby destroyed, and the cylinder containing the hydrogen gas will rise.

Experiment B.—Suspend the paper cylinder to the balance *mouth upwards* and counterbalance it; then pour into it from a jar a quantity of carbonic acid gas. This will cause the paper cylinder to descend, because carbonic acid gas is heavier than atmospheric air in the proportion of 3 to 2.

The two jars of gases should be provided beforehand.

If time permits, the gases may be introduced into the paper cylinder by the tubes coming directly from the generating gas bottles. But this method is slow and uncertain.

Price of the paper cylinder, mounted with strings, 6d.

654. *Balloons*, for rising in the air when inflated by hydrogen

gas, or coal gas. Spherical shape, commonly coloured in gores; but the 9, 10½, and 12 inch sizes can be supplied coloured plum-pudding pattern.

9 inch, price 1s. 3d.
10½ inch, price 1s. 9d.
12 inch, price 2s. 6d.

15 inch, price 3s.
18 inch, price 5s.
6 inch, for hydrogen gas, price 1s.

Balloon, in shape of Mr. Punch, 6 feet high, 9 feet in circumference. Price 63s.

Balloon, in the shape of an elephant, 3 feet long, 2½ feet high, ascends in proper position, and can be led by a guiding string into any part of a room. Price 35s.

Balloon, fish shape, 15 inches long. Price 4s.

These balloons are made of gold-beater's skin. When not in use they should be kept in a clean tin case, with a little camphor, to preserve them from the attacks of insects. They ought never to be wetted. When they are expanded with air from the mouth, the lips ought not to touch the mouth of the balloon. They will all ascend with dry coal gas. In filling a balloon the gas should be passed through a glass tube containing pieces of chloride of calcium, to free the gas from water. The large sizes of balloons, when in good condition, may be kept suspended two or three days, fastened by a string to the table; but the hydrogen gas is gradually exchanged for atmospheric air by osmotic action, and the balloon finally descends. The hydrogen gas with which a balloon is to be filled must all be prepared before the filling is attempted, and should be contained in a gasholder, or in large jars provided with stopcocks.

655. *The Baroscope, or Balance, with weight and cork,* to show how the bulk of objects affects their weights when they are weighed in atmospheric air, and in vacuo. *Price without the receiver,* 6s.

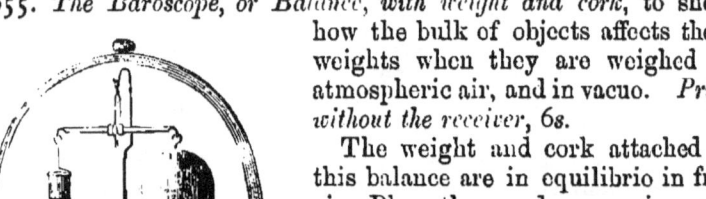

655.

The weight and cork attached to this balance are in equilibrio in free air. Place them under a receiver and exhaust the air. The cork will then appear to be heavier than the weight; for as its bulk is greater than that of the metal of which the weight is made, it must be more sustained by the air. Admit air to the receiver, and the original equilibrium is restored.

For, as bodies which are immersed in liquids are known to lose a part of their weight, equal to the weight of a quantity of the

EXPERIMENTS ON THE RESISTANCE OF AIR. 141

liquid of the same bulk as the immersed body, so bodies of different specific gravities, which are in equilibrio in air, will not be so in vacuo; for then they will gain that weight which they lost in air, and the body of the greatest bulk will gain the most.

As cork changes in weight when exposed to moist air, it is necessary previous to every experiment to justify the counterpoise, which is made to open for that purpose.

A receiver of the form of fig. 825 is generally used for this experiment, because the receiver of fig. 655 requires a very wide plate to rest on.

656. *Water Hammer, or Philosophical Hammer,* for showing the force and solidity with which water falls in a vacuum.

This is a glass tube, partly filled with water, and quite freed from air. When this is held vertically in the hand, and smartly jerked up and down, the water falls in a block, with a noise resembling that produced by the stroke of a hammer. The most effective motion is a steady lift upwards, succeeded by a powerful and sudden jerk downwards. If the jerk is too strong, the water sometimes drives out the bottom of the tube.

Several sizes are supplied:
7 inch, *price* 1s. 6d.; 9 inch, *price* 2s.; 12 inch, *price* 2s. 6d.

656 A.—*Water Hammer,* V-formed, about 16 inches long, closed at one end and mounted at the other end with a brass stopcock, suitable for exhaustion by the air-pump, fig. 656 A. *Price, with stopcock,* 8s.

Management of the V-formed Water Hammer.—Fig. 656 A represents the apparatus in four positions. It consists of a V-shaped tube, having a brass collar fixed on its open end and joined to a stopcock. There must be a leather washer between the two brass pieces, and that washer must be in good condition, well oiled, soft and flexible. One limb of the tube is three-fourths filled with water.

a. Hold the apparatus in this position, screw the stopcock into the plate of the air-pump, and exhaust thoroughly. The air will be seen to rush out of the water for a long time. When the discharge of air appears to cease turn off the stopcock, unscrew the apparatus from the pump, and transfer the water to the branch over the stopcock.

b. Then hold it in this position. If there is any leakage about the brass fittings, either in the plug of the stopcock, or in the cement, or about the washer, air will rush up in bubbles through the water.

142   PNEUMATICS.

Correction of the leakage must be made and the air be again exhausted, until the water remains quiet in the position marked *b*.

c. Transfer the water to the closed limb and hold the apparatus in position *c*, applying your hand to the stopcock. Give the apparatus a jerk up and down, as directed in § 656, when, if the exhaustion is good, the water will fall like a solid block. If the exhaustion is not good, then upon tapping the table with the bottom of the tube that contains the water a slight jingling noise will be heard, and a multitude of minute bubbles of air will rise up through the water.

d. Put the tube into the position marked *d*. If the exhaustion is good, the water will remain in the closed branch, as represented; but this will require very complete exhaustion.

This experiment affords an excellent illustration of the difficulty of completely abstracting air from water.

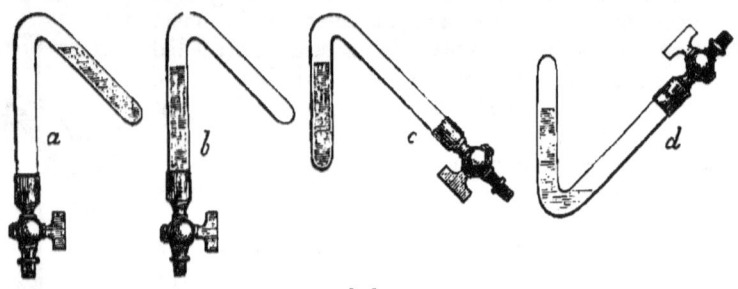

656 A.

656 B. *Fork-shaped Water Hammer*, fig. 656 B. Price 2*s*. 6*d*.

657. *Guinea and Feather Experiment*, to show that heavy and light articles fall with equal rapidity when air is removed.

658. *Brass-work for this Experiment*, consisting of a 4-inch ground brass plate, with stuffing-box, sliding rod, and two falling stages; the plate is also furnished with a hook, on which to hang a bell, the Magdeburg hemispheres, &c. See fig. 660. Price 18*s*.

659. *Brass-work for three falls.*—Similar brasswork arranged for three falls, but not provided with the hook. Price 21*s*.

660. *Open Conical Glass Receivers for the Guinea and Feather Experiment*, flanged at both ends and well ground.

656 B.   See sizes and prices at Nos. 888, 889, 890.

664. *The Guinea and Feather Apparatus* is intended to illustrate the fact that the air diminishes the velocity of falling bodies, and this diminution is greater or less according to the

EXPERIMENTS ON THE RESISTANCE OF AIR. 143

density of the falling body; but if the air is removed, the velocity of all falling bodies is equal, whatever difference may exist in their comparative densities. To show this with effect, place a tall open receiver, No. 660, on the plate of an air-pump, and on the top of the receiver put the guinea and feather apparatus, having previously placed on the flap of the apparatus a guinea or other coin, and a feather; exhaust the receiver, then turn the milled-head until the flap falls, and the guinea and the feather will strike the plate of the air-pump at the same time. But if the experiment is repeated without exhausting the receiver, the resistance of the air will cause the feather to occupy a much longer time in descending than the guinea.

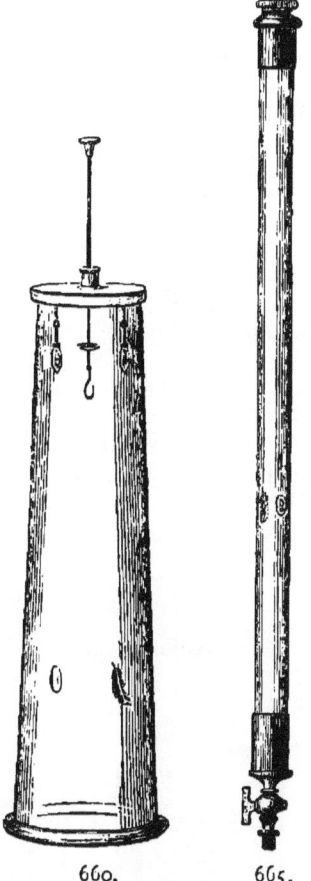

660.   665.

The eye of the observer should be directed towards the bottom of the receiver, to ascertain accurately the coincidence of descent in the two bodies.

Guineas are never used in this apparatus, and feathers not always. Common coins, farthings, or even disks of lead, answer the purpose, and very often round pieces of thin paper are used instead of feathers. The latter are liable to stick to the sides of the receiver, and to be thus arrested in their fall.

665. *The Guinea and Feather Experiment* can be performed with a *glass tube;* size about 4 to 5 feet long and about $1\frac{3}{4}$ inches wide, mounted with brass caps at both ends, and with one brass stopcock, having two male screws. Fig. 665. Price 15s.

666. *Windmill*, with two sets of equal-sized vanes, one set striking the air edgeways, the other set striking it with their broadsides. In the open air the latter soon become still, while the former continue to revolve. In vacuo they both travel the same length of time. Fig. 666. Price 36s.

To set the mills in motion in the open air, pull up the rod in the middle, which puts both sets of vanes in action. The broadside vanes ought to stop some time sooner than the set fixed to run edgeways. If they do not, they are out of order; most probably some object is adhering to one of the sails, making it heavier than the others, and therefore irregular in action. In that case, if you turn the mill slowly you will find the stoppage to occur whenever one particular sail, after a little oscillation, rests lowest. That points out where the correction is required.

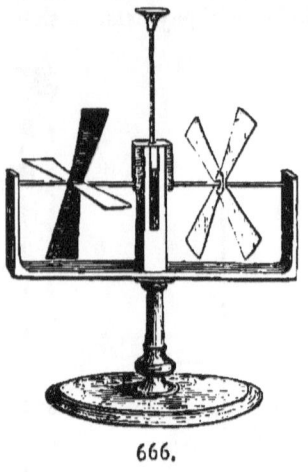

666.

If the mills work regularly in the open air, put them under a receiver of the form of fig. 696, mounted with the sliding rod, fig. 625, so arranged that the lower end of the sliding rod is put above the knob of the windmill, when that knob is pulled up as far as it will go, which it should be before being placed under the receiver. The mills should be placed on a suitable grooved plate. See No. 641. When the receiver has been sufficiently exhausted, the rod No. 625 should be suddenly pressed down and the mill be set in action. If in good order, the two sets will remain in motion during the same time.

GROUP B.—ON THE EXPANSION OF AIR, § 671 to § 710.

671. *Bulb Gauge and Glass Jar, for testing the exhausting power of an Air-pump;* the bulb about an inch in diameter and the stem about 4 inches long, in a jar of about 5 ounces capacity. *Price of the pair 2s.*

671.

Experiment A.—Place the apparatus (after putting a little water into the jar) upon the plate of the air-pump, cover it with a receiver, and exhaust the air. The pressure being removed from the surface of the water in the jar, the air in the bulb will expand and escape in large bubbles through the water. After sufficient exhaustion, admit the air into the receiver, and its pressure on the water in the jar will force it up into the bulb, leaving therein only a small bubble of air. The bulk of this bubble, compared with the bulk of the tube and bulb, indicates the degree of the exhaustion.

EXPERIMENTS ON THE EXPANSION OF AIR. 145

Experiment B.—Fill the bulb-tube almost full of water, leaving only a very small bubble of air. Fill the jar with water, invert the tube and plunge it into the jar; then remove most of the water from the jar by means of a syphon or pipette, leaving enough to cover the mouth of the bulb-tube. Place the jar under a receiver and exhaust it. The small bubble of air will then expand and force all the water out of the bulb-tube. On readmitting air to the receiver, the water will be driven back from the jar into the bulb-tube.

672. Experiment C.—Similar experiments to A and B can be made with vessels prepared for other purposes, but combined for this experiment, as represented in fig. 672. *a* is the plate of the air-pump; *b* a guinea and feather glass, serviceable also as a fountain glass; *c* a ground glass plate to close it; *d* a tall jar on foot, resting in this case on a small porcelain support (see No. 641); *e* a bulb-tube with long stem. For the experiments just described the figure shows too much water; but that is subject to correction.

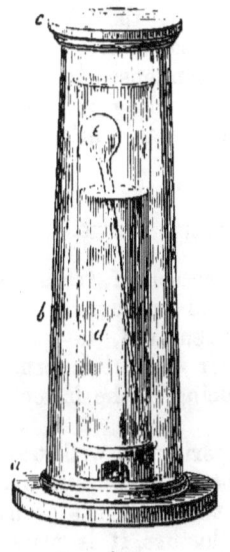
672.

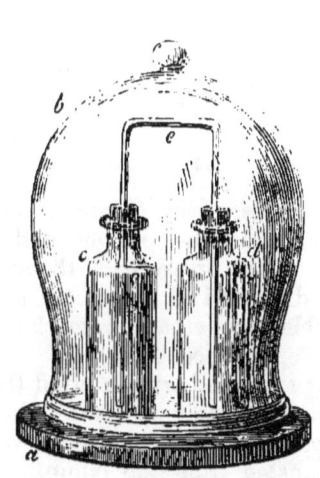
673.

673. *Pair of narrow Bottles*, connected by a bent tube fastened by a cork to the mouth of one of them. *Price* 1s. 6d.

Fig. 673 *a* is the plate of the air-pump, *b* a receiver. The bottle *c* is supposed to be *nearly* full of water, but retains a small quantity of air. The cork must fit tight. When the air is exhausted from the receiver, the expansion of the small quantity of it left in *c* drives the water through the tube *e* into the bottle *d*. When air is

L

readmitted into the receiver *b*, the water is driven back from the bottle *d* into the bottle *c*.

674. *Lungs Glass*, fig. 674. *Price, without the receiver*, 5s.

The apparatus with this fanciful name is constructed as follows:—

674.

A globular glass on foot is surmounted by a brass cover. A bladder is tied round a small pipe that descends from the cover into the globe, but the air within and without the bladder is not in communication. When this apparatus is placed under a receiver, and the air is partly withdrawn from the bladder, the spring of the air surrounding the bladder in the globe, which cannot escape, *compresses* the bladder; when air is let into the receiver, the bladder *expands*. That is the whole action.

These motions are slow when held to be analogous to those of the lungs. The best way to proceed is as follows:—Partially open the aircock that supplies the receiver with air, and leave it in that state, and then exhaust slowly. Every stroke of the exhausting piston partially condenses the bladder, but the aircock meanwhile restores the expanding pressure. A little experience suffices to enable the operator to work the apparatus with some regularity.

EXTRICATION OF AIR FROM VARIOUS BODIES.

675. *Air from Spring-water.*—Place a vessel of fresh spring-water under the receiver of the air-pump and exhaust the air. As soon as the pressure of the atmosphere is removed, the gases that are dissolved in the water expand by their elasticity, form large bubbles, and escape from the water, producing in the latter a state of effervescence.

It must be borne in mind that in all experiments with water the vapour of water passes into the works of the air-pump, and makes it impossible to procure a good vacuum until the water has been again exhausted from the pump. Hence, at a lecture, it is prudent to avoid experiments with water until all other experiments that demand a good vacuum have been performed.

676. *Air from the pores of Plants.*—The plant selected may be a sprig of fresh mint, but many other plants serve equally well, either fresh or dried. Fasten the stalk of the plant in the clip No. 626, and lower the clip and weight by a wire or cord to the bottom of a tall jar containing fresh and clean spring-water. Place the jar on the air-pump plate and cover it with a tall receiver. The arrangement is shown in fig. 676. A shorter and wider apparatus may also

be used. Exhaust the air from the receiver. The whole surface of the mint will be covered with minute bubbles of air, which give it a beautiful appearance. These are disengaged in part from the spring-water, and in part from the plant.

Price of tall receiver, 6 inches diameter and 20 inches high, 25*s*.

Price of tall jar, 4 inches diameter and 16 inches high, Bohemian glass, 8*s*.

677. *Air extricated from Cork.*—Fix some lead to a piece of cork, enough merely to submerge it in water. Place the jar containing the cork and water under a receiver, and exhaust by the air-pump. The air contained in the pores of the cork will then expand, so as to swell it and make the mass lighter than water. It will consequently rise to the top of the water and swim, cork and lead together, and the surface will be covered with a beautiful mass of globules of air, resembling the pearly drops of dew on the blades of grass. When the air is admitted into the receiver, the bubbles of air disappear and the cork sinks.

678. *Air from Beer or Ale.*—Place a beaker, No. 678, partly filled with beer or ale, under a receiver, and exhaust. The liquor will expand, and a mass of froth rise up in the beaker to form a large white head. This production of froth is due to the tenacity of the fluid, which prevents the bubbles of air from escaping as soon as they rise. When the air is readmitted to the receiver, the bubbles contract and subside. On tasting the beer, after this operation, you will find it to be flat and spiritless.

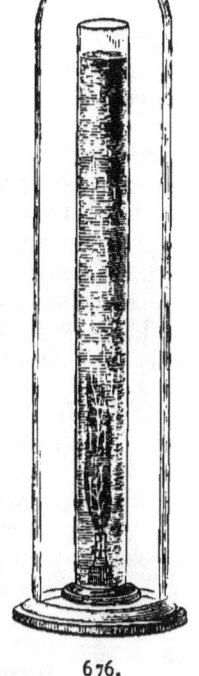

676.

679. *Air from an Egg.*—If you put an egg into a beaker of water, place it under a receiver, and work the pump to produce a vacuum, you will see the air rise in a very beautiful manner in small jets through the water from the pores of the egg-shell.

678.

680. *Air from Coke.*—Put two pieces of coke into water in a beaker; both will swim. Tie one of the pieces to the weight and clip No. 626, and sink it to the bottom of some water in a beaker. Put the glass under the air-pump receiver and exhaust. A great quantity of bubbles of air will come from

the submerged piece of coke. When the discharge of air slackens, take the coke from the clip and add it to the other piece of coke still swimming in the original beaker, in the water of which it will instantly sink to the bottom, showing that water, when present in coke instead of air, greatly adds to its weight and specific gravity.

681. *Elasticity of the Air-bubble in an Egg.*—The air-bubble in an egg is situated at the great end between the skin and the shell. On making a small hole in the little end of the egg and inverting it in a glass, and placing the glass under an air-pump receiver and exhausting it, the bubble of air will expand and force out the contents of the egg through the small hole in the lower end.

682. *Fruit and Taper Stand,* fig. 682. *Price* 3s.

Shrivelled apples, prunes, or raisins, after being soaked a little in warm water, when placed under the receiver of an air-pump, on this stand, will become plump on the receiver being exhausted, from the elastic force of the air within their cells; but on readmitting the air the fruit loses its apparent freshness and re-assumes the same appearance as before the experiment.

This stand also serves to support a taper under the air-pump receiver, to show that it will not burn when air is removed.

682.

683. *Apple-cutter,* fig. 683. *Price* 3s.

This is a glass jar mounted with a brass collar provided with a circular cutting edge. Upon this circular knife an apple is fixed by gentle pressure, care being taken to see that the apple leaves no part of the cutting edge free. The jar is then placed on the plate of the pump, and the process of exhaustion is commenced. As this proceeds, the apple descends under the continued action of the circular knife and the pressure of the atmosphere, and a cylindrical mass falls into the jar.

683.

684. *Bladder experiments,* to show the expansive power or elasticity of the air.

Take a clean and soft bladder, containing a little air, and tie the neck close; then suspend the bladder by a hook, similar to No. 624, under a receiver, similar to No. 684; place the receiver on the plate of the air-pump, and exhaust the air. Thereupon the bladder will be expanded and blown up by the spring of the small quantity of air left in it. See fig. 684. On re-admitting air to the receiver, the bladder will shrink into its former shape.

685. *Price of bullock's bladders, prepared for use,* 6d.

686. *Sheep's bladders, prepared for use,* 2d.

688. Bladders should be occasionally washed with a mixture of

water and glycerine, to prevent their cracking into holes. After using a bladder it should, while still moist, be fully blown up and then sponged with a mixture of glycerine and water. Before submission to an experiment at a lecture, the soundness of bladders should be ascertained by trial; for they are constantly subject to perforation by insects.

Solution of carbolic acid in water deprives bladders of a bad smell when they are put aside in a wet state.

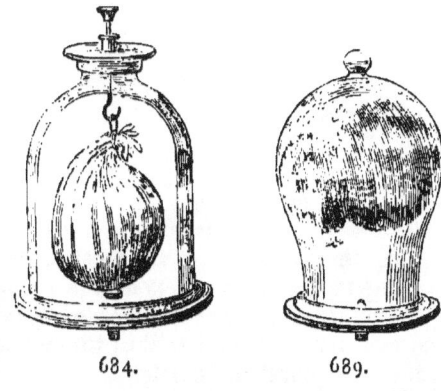

684.   689.

689. *Elastic Balls.*—Four caoutchouc balls containing a little air, and sealed air-tight by dipping their mouths into chloroform, for showing the elasticity of air by expansion in an exhausted receiver. *Price of the set* 1s.

A single caoutchouc ball, firmly closed and containing a little air, can be expanded under a receiver until it becomes very large.

Fig. 689 represents a curious experiment. Two sealed caoutchouc balls, containing a little air, are put under a small bell-shaped receiver, which is exhausted. As the action proceeds the balls swell and soon become too large to lie quietly together on the plate, whereupon a strife takes place between them and they both fly to the upper wider part of the receiver and remain there, side by side, resting against the walls of the jar.

Consult § 628 for another experiment with caoutchouc balls.

690. *Bladder frame and lead weights,* to show the elasticity of air when relieved from pressure, figs. 690 and 691.

690. Small size, diameter of lead weights, $2\frac{1}{2}$ inches. *Price* 8s.

691. Large size, diameter of lead weights, $3\frac{1}{2}$ inches. *Price* 16s.

The bladder used for this experiment must be of small size, sound, contain

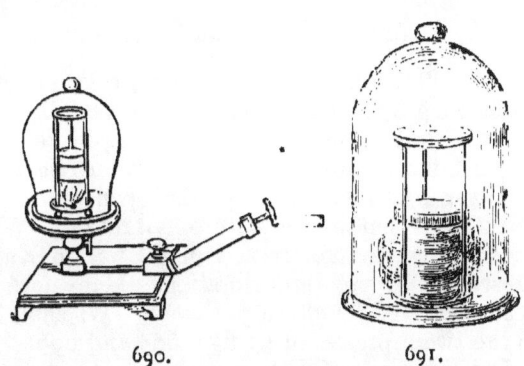

690.   691.

a little air, and be well tied. It should also

have been rendered flexible by soaking in water, mixed with a little glycerine.

This apparatus illustrates the expansive force of the air. The small portion of air within the bladder expands on removing the external pressure, by means of an air-pump and receiver. If the exhaustion is continued, the expanded air will exert such force as to raise the heavy weights placed upon the bladder. On admitting the air the bladder shrinks to its original bulk, and the weights descend.

692. *Glass breaking squares*, to show the expansive power of the air. *Price each* 1s. 6d.

These are thin square glass bottles, of about half a pint capacity. To use them take one of the bottles, cork it, and cover the cork with sealing-wax, so as to prevent the escape of the enclosed air; then set it under the receiver of an air-pump and exhaust the air from the receiver. The air within the bottle will then expand with such force as to burst the bottle.

The wire cage, § 693, is required for this experiment.

693. *Wire Cage*, to cover the breaking squares and collect the fragments of glass when they burst. *Price* 3s.

It is prudent also to cover the plate of the pump with a thick piece of paper, or a disk of felt, to prevent scratches from the fragments of the exploded glass.

694. *Paper Smoke-jacks*, two patterns, a wheel and a serpent. *Price of the pair* 1s.

They serve to show the rising currents produced when air is warmed and expanded. They may be suspended over a lighted lamp by a thread, or on a needle point, placed on the side of a magnet. The vanes of the wheel must be bent when the apparatus is to be used.

693.

Bells to be rung in Exhausted Receivers.

695. Bell suspended in a brass circle, mounted on a wooden foot, to be rung by being shaken. *Price* 5s. 6d.

696. Bell mounted on a brass foot, to be rung by a blow on the upper table, mounted on a felt disk. *Price, without receiver*, 5s. 6d.

When a bell is struck in the open air it *rings*, because the air about it is put into a state of vibration. When a bell is struck in an exhausted receiver it ought *not* to ring, because no air is present to be put into vibration. Many forms of apparatus have been contrived for this experiment. We shall limit our description to the two represented by figs. 695 and 696.

695 Unscrew the brass-work from the mahogany foot and screw

EXPERIMENTS ON THE EXPANSION OF AIR. 151

it in the centre of the air-pump plate. Cover it with a receiver, lift up the pump and *shake it*. This last direction brings a difficulty. If you use Tate's pump, or any of those from No. 503 to No. 521, you *cannot shake it*. It is only with small pumps such as Nos. 522, 523, that you can put this process into operation. The shaking makes known another difficulty. If you have a small receiver over the bell, the shaking produces scarcely any sound, although the receiver is full of air. The reason is, that in a small receiver there is not space enough for sufficient air to vibrate. The difference in the sound of the bell when shaken in the open air and when shaken in a small receiver is very great. However, you proceed to exhaust the receiver as completely as you can, and then shake again. You will still hear some sound, because there is a continuous metallic contact between the bell and the open air through the plate of the air-pump. The difference in the sound of the bell when thus rung, first without exhaustion and then after exhaustion, is so slight, that the experiment is a failure. If you use Tate's air-pump, the only way to succeed with this experiment is to use an extra pump-plate, No. 563, and the largest receiver you have at command, that is not too wide for the pump-plate.

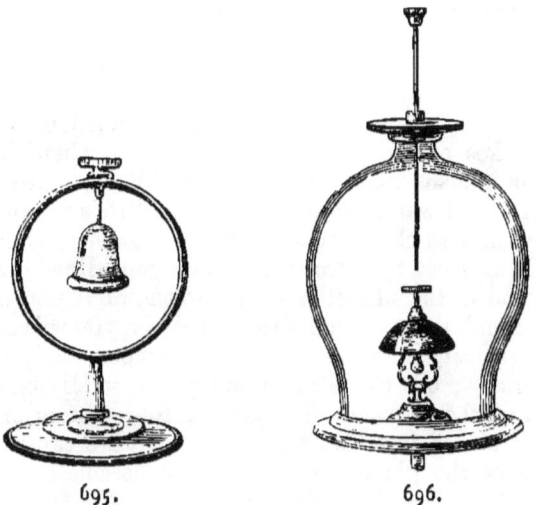

695.   696.

696. This apparatus represents a call bell, mounted on a brass foot which rests upon *a thick disk of felt*. The bell is rung when a blow is struck upon the small table which surmounts it. A bell-shaped receiver is put over it, the largest at command, and the widest that fits the air-pump table. The receiver has a neck and ground mouth, upon which is placed the brass plate and sliding-

rod, fig. 625, the hook at the bottom of it being first removed. The apparatus, as represented by fig. 696, being arranged upon the plate of the air-pump—the receiver being greased, if necessary, to make it fit the table—a blow is given to the bell by forcing the sliding rod down with a slight jerk. The bell sounds with an intelligible note. The receiver is then to be exhausted as thoroughly as possible, and another blow is to be struck by the sliding rod on the bell. If the exhaustion is good, there will be scarcely any sound produced. The thick disk of felt is interposed between the bell and the pump-plate to prevent the direct conduction of the sound through the pump. Fig. 696 may be called a successful bell experiment.

700. *Leslie's Apparatus for freezing water in vacuo*, over oil of vitriol, consisting of a flat bell-shaped receiver for the air-pump, a porcelain pan for the sulphuric acid, and a glass capsule for the water.

The apparatus is represented by $j\ k$, in fig. 518. *Price of the set 7s. 6d.*

701. When the apparatus is to be used there should be put into the pan, half an inch in depth, of the strongest oil of vitriol (*not* the fuming Nordhausen acid). The pump must be well screwed up, the edge of the receiver be greased, to make it fit close, and the whole apparatus should have been kept for some time in a cool place. *The above apparatus is suitable for use with Tate's small air-pump.*

702. This experiment succeeds best when the pump, the water, and the air of the room are as cold as possible. The pump and every part of the apparatus being in good condition, it takes twice as many strokes of the piston to freeze water when the air is at 60° Fahr. as it takes when the air is at 40° Fahr. In winter, when the apparatus and water are tolerably cold, Tate's pump will freeze the water with less than 100 strokes. In summer, at about 60°, it will require at least 150 strokes; and if you allow the water and pump to stand in the sun till they are warm, or if you take diluted acid, or too much water, or too large a receiver, you will fail entirely.

A small quantity of water is, of course, more easily frozen than a large quantity; but when Tate's pump (the small size) is in good condition, and the weather cold, three or four ounces of water can be frozen, if placed in a thin porous earthenware capsule.

In all cases the glass receiver must be as small and as flat as possible.

703. *Porous evaporating Basins*, for use in freezing large quantities of water, when the more powerful pumps are used; red clay, *Wedgwood's.*

4 inches diameter, *price* 8d.
5 inches diameter, *price* 1s. 4d.

EXPERIMENTS ON THE EXPANSION OF AIR. 153

704. It is often recommended to freeze water by the evaporation of sulphuric ether in an exhausted air-pump receiver. That operation has several disagreeable points. The ether is very expensive, the vapour from it fills the whole room and is especially unpleasant to the operator; it attacks the oil in the air-pump, and it fails more frequently than the process with sulphuric acid, when the latter is made with the precautions just pointed out.

705. *Evaporation in a Vacuum.*—It is often desirable in chemical operations to expel water from substances which cannot be heated without being made to undergo changes that are injurious. In such a case the chemist avails himself of the power of the air-pump. A porcelain pan, similar to fig. 705, is half filled with concentrated sulphuric acid, a liquid which rapidly absorbs aqueous vapour, and this is placed upon the table of the air-pump. The substance that is to be evaporated or dried is placed in a watch-glass or a porcelain capsule upon this pan. The whole is covered with a flat glass receiver, such as is represented in fig. 518, and the air is exhausted from the receiver. Water then rises from the substance that is to be dried to form an atmosphere in the receiver, but the concentrated sulphuric acid rapidly absorbs it; fresh vapour is then formed and is absorbed by the acid, and thus the evaporation proceeds till the required result is produced.

It is sometimes expedient to dry substances contained in filters without removing them from their glass funnels; in which case the arrangement represented by fig. 706 is useful. The lower part of this apparatus is made of porcelain, and the upper part of wood. Acid is put into the porcelain pan, and the wooden table being placed over it, the funnels and capsules containing the mixture that is to be dried are placed upon the table. The exhaustion and evaporation then proceed as described in the preceding paragraph.

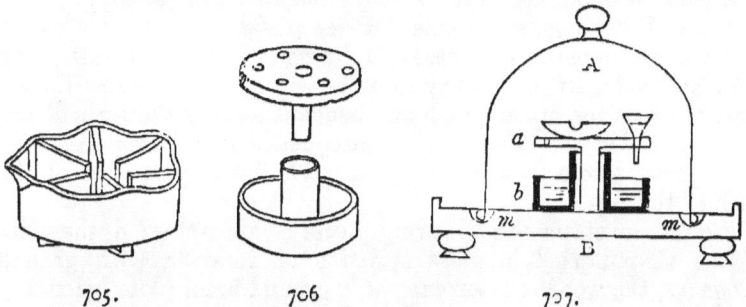

705.   706.   707.

The apparatus last described is sometimes employed with concentrated sulphuric acid to dry substances without the aid of the air-pump. The method of proceeding is shown by fig. 707. B is a round wooden table, in which is turned a circular groove, m m. A is a glass or metal receiver, which fits the groove m m. a and b represent the apparatus, which is shown separately by fig. 706. If the receiver A is made of tin-plate, the groove m m is filled with oil. If the receiver is of glass, the groove may be filled either with oil or mercury. With this apparatus the evaporation goes on more slowly, because the air in the receiver prevents the rapid vaporisation of the water.

GROUP C.—ON THE PRESSURE OF THE AIR, § 711 to § 745.

711. *Downward pressure of the Air.*—Place a receiver on the plate of the air-pump and exhaust the air as completely as the pump permits. A gauge, § 575, may be screwed into the plate under the receiver, to show the degree of exhaustion attained. The receiver will become so firmly pressed down upon the plate by the external air, that it would be exceedingly difficult to separate it by force. This will not appear surprising when we consider that the glass is pressed down with a force equal to as many times 15 pounds as there are square inches covered by the bottom of the receiver. That is to say, a receiver of 7 inches diameter is pressed by about 577 lb., one of 8 inches by about 754 lb., and one of 10 inches by about 1178 lb. Hence the necessity, before removing an exhausted receiver, of relieving its pressure by an admission of air through the aircock.

711.

712. *Filter Cup*, for causing mercury to be forced by the pressure of the atmosphere in a copious shower through extremely fine pores in a piece of wood, fig. 712. *Price of the filter cup 7s. 6d.*

712 A. Filter cup, small size. *Price 4s. 6d.*

Place the apparatus represented by fig. 712 upon the air-pump plate and put a little mercury into the cup *c;* then exhaust the air from the receiver, upon which an abundant silvery shower will fall from the end of the filter cup, in consequence of the pressure of the atmosphere forcing the mercury through the empty pores of the neck of the cup.

Fig. 712 contains the following pieces of apparatus: *a*, the plate of the air-pump; *b*, a glass cylinder or receiver, with ground flange; *c*, the wooden filter cup; *d*, a ground brass plate belonging to the filter cup; *e*, the ground neck of the glass receiver; *f*, the porous neck of the filter cup; *g*, a glass jar, set to catch the mer-

EXPERIMENTS ON THE PRESSURE OF THE AIR. 155

cury and prevent its entering the pump; *h*, a small porcelain stand, to keep open the way to the exhausting power. This is a complicated arrangement for a simple experiment; but it serves to show one of the ways to be used to keep mercury, water, &c., from entering the air-pump.

713. *Shower of Air in water, Apparatus for showing*, fig. 713. Price 7s. 6d.

This apparatus consists of a filter cup, the neck of which is a porous cane. When this is arranged as shown by fig. 713, and the receiver is exhausted, the pressure of the atmosphere through the cane and water to fill the vacuum in the receiver causes a copious shower of air to appear in the water.

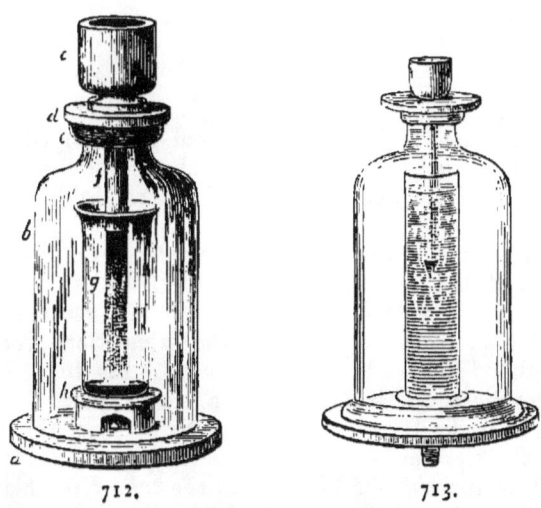

712.   713.

714. *Hand and Bladder Glass.*—This instrument has this double name, because it is used for two purposes. There are also two forms, which differ in quality of glass and in price.

714. Conical form, German glass, fig. 714. *Price* 1s. 6d.
715. Swelled form, fine Bohemian glass, fig. 715. *Price* 3s.

When placed with the narrow end uppermost, they are called hand glasses. When placed with the wide end uppermost, they are called bladder glasses. They must both have strong welts and be well ground on the edges. Their uses are as follow:—

716. *Experiment with the hand glass.*—The hand glass is used to prove the pressure of the atmosphere. Set the large end on the plate of an air-pump, and place the palm of the hand so as to cover and close the small end. On the glass being exhausted, the pressure of the air is felt so strongly that it is nearly impossible

to move the hand; but if the air is admitted into the receiver, the hand will be again loosened. When this experiment is made in a school, the narrow end of the cone should be so small as to be completely closed by the small hand of a child, otherwise the experiment fails.

714.   715.   716.

717. *Experiments with the bladder glass.*—Preparation: a piece of thin and clean pig's bladder, soaked in water till it is quite flexible, is to be tied tightly over the wide end of the bladder glass and set aside till it is become dry and drawn tight, as represented by fig. 715.

The narrow end of the glass is to be placed on the plate of the air-pump, and the air to be exhausted from it. The bladder, yielding to the external pressure of the air, assumes a concave figure, which increases in depth as the exhaustion proceeds, until the strength of the bladder is overcome by the incumbent weight of the atmosphere, and it bursts with a great report.

In several books it is recommended to perform this experiment on the pressing powers of the atmosphere by placing a thin plate of window glass, instead of a bladder, on the top of the bladder-glass. If this is done, the glass is no doubt broken with a loud report; but I have also known the glass cone to be broken and the glass to be scattered about dangerously. This experiment includes also the special chances of damaging the surface of the plate of the air-pump by scratches effected by the glass fragments, and of driving fragments of glass into the body of the pump through the hole in the plate.

Immediately before use it must be ascertained that the bladder is dry and well stretched. The safest plan is to dry the bladder before the fire at the time. If it is left damp and slack, the experiment fails. If a barometer gauge is attached to the pump, access to it should be cut off by closing its stopcock, or it should be altogether removed from the pump before this experiment is made; because the sudden rush of air into the pump, when the concussion occurs, is liable to destroy the gauge.

EXPERIMENTS ON THE PRESSURE OF THE AIR.   157

718. In considering the performance of experiments with the hand and bladder glass, it is proper to bear in mind the approximate weight of the atmosphere pressing upon openings of different sizes. Thus:

If the diam. of the glass is 1½ in., the pressure is about 26 lb.
„   „   „   2 in.,   „   „   47 lb.
„   „   „   2½ in.,   „   „   73 lb.
„   „   „   3 in.,   „   „   106 lb.

This sufficiently accounts for the bursting of a pane of glass, or a dry and strained slip of bladder. But it also warns the operator to be cautious in experiments with the hand glass, and especially to see that the rim of glass on which a child is desired to press his hand is suitably prepared for that purpose, that the glass is thick and smooth, and free from sharp edges or points.

719. *Weight of the Atmosphere.*—Apparatus, consisting of a square bar of iron, 1 inch diameter, 15 lb. weight, to show to young people the amount of the pressure of the atmosphere *per square inch* at the surface of the earth. Price 6s.

The length of bar iron having a cross section of 1 square inch, proper for this model, is ascertained as follows: 15 pounds multiplied by 7000 gives 105,000 as the corresponding weight in grains; 252·458, the number of grains in a cubic inch of water, multiplied by 7·7, the specific gravity of bar iron, gives 1944 grains as the weight of a cubic inch of iron. Then 105,000, divided by 1944, gives 54 as the number of cubic inches of iron demanded. But the bar itself being 1 inch square, it is only necessary to cut off a piece 54 inches in length, to have a measure of the weight of the atmosphere. See § 752.

720. *Crushing Power of Atmospheric Pressure*, exhibited by a cylinder of thin tin-plate, measuring about 10 inches high by 4½ inches wide. Fig. 720. Price 1s. 6d.

A small quantity of water is boiled in this vessel over a gas-burner or a spirit lamp

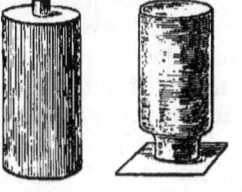
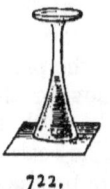
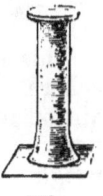

720.   721.   722.   723.

till the atmospheric air is expelled by the steam. The cylinder is then promptly closed by a good-fitting cork and is placed in a pan, and a quantity of cold water is poured over it: the cylinder immediately collapses.

721. *Upward Pressure of the Air.*—Use for this experiment a glass jar with a smooth and level mouth. The form and size of the

jar are immaterial. Those represented by figs. 721, 722, and 723, all answer the purpose. Fill the jar quite full of water, cover its mouth with a piece of Bristol board or stiff writing-paper, put one hand lightly but steadily on the paper to keep it close to the glass, and with the other hand quickly invert the glass jar. It will then remain quite full of water, as shown by the figures, even when the hand is removed from the paper. The upward pressure of the air prevents the escape of the water as long as the glass jars are steadily kept in a vertical position. It is upon this principle of the effect produced by the upward pressure of the air that pipettes act (see § 361), and that mercury is upheld in the barometer (see § 747).

Prices of these glasses:—

721. Size 6 by 3 inches, mouth 2½ inches, 1s.
722. Tall conical test glass, 3 inch mouth, 1s.
723. Jar with foot and flange, 6 by 2 inches, 9d.
724. *Bottle of japanned tin-plate, with perforated bottom,* to show the upward pressure of the air. *Price 2s.*

To fill the bottle dip it into a deep beaker filled with water, permitting the air to escape from the bottle by the hole in the neck. When the bottle is full, close the hole by pressing a thumb on it, and lift the bottle out of the water, keeping it upright. Then, although the bottom of the bottle is full of holes, the water is retained until the hole at the top is opened. In short, this bottle acts on the same principle as the pipette. See No. 361.

725. *Syringe and Lead Weight,* for showing the pressure of the atmosphere. *Price, without the receiver and hook,* 10s.

724.   725.

This apparatus is intended to prove the pressure of the atmosphere. If the piston is pulled out to the fullest extent in the open air, on removing the acting force it immediately resumes its usual position, although the lead weight may be so placed as to exert its weight. The reason of this is, that when the weight is pulled down a vacuum is produced in the syringe, and there is no valve to permit the supply of air. But if the apparatus is suspended by the handle from the inside of a receiver placed on the plate of an air-pump, and the air is exhausted, the weight will immediately begin to descend, and its descent will be proportionate to the amount of exhaustion. On re-admitting the air to the receiver, the weight

EXPERIMENTS ON THE PRESSURE OF THE AIR. 159

and body of the syringe will ascend in consequence of the atmospheric pressure.

726. *Set of three Glass Globes,* fixed one over the other, and so connected as to show the pressure of the atmosphere; size of the globes 3 to 3½ inches diameter, with brass connectors. *Price of the set* 16s.

727. The set mounted with corks instead of brass tubes. *Price* 8s.

728. The glass receiver, to cover them, size 6 inches diameter, 20 inches in height. *Price* 25s.

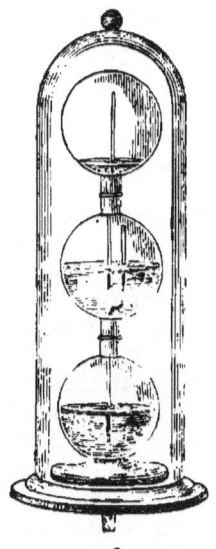

728.

Use.—Nearly fill the lowest globe with coloured water. The brass work unscrews between the lowest and the middle globe for that purpose. Set the apparatus on the air-pump plate, cover it with a tall receiver, and exhaust. A small hole is pierced near the top of the middle globe, through which the air of the two upper globes escapes during the exhaustion of the receiver. The confined air left in the lowest globe will then expand and drive part of the water up into the middle globe through the jet pipe which extends from the bottom of the lowest globe to the top of the second. Admit air into the receiver; that air will enter the middle globe by the small hole pierced in it, and its pressure will force up a quantity of water from the middle globe into the upper globe. After the experiment, the globes are to be separated by unscrewing the brass fittings, and emptied.

729. *Fire Syringe,* to set fire to German tinder by the sudden compression of air in a brass syringe. *Price* 4s.

The German tinder should be quite dry, and a small piece of it be put into the cavity at the end of the piston. The piston should be pushed into the

729.

tube, and the knob at the far end of the tube being placed against some firmly-fixed object, the piston should be suddenly and forcibly driven in, and immediately withdrawn, when it ought to bring with it the lighted tinder. The operation requires a little practice.

729.

730. *Bottle Imps, or Cartesian Devils*, to illustrate the compressibility and elasticity of air.

730. Small grotesque figures of coloured glass, a variety. *Price, each* 6d.

731. Larger figures, gondola, balloon, cage, &c. *Price, each* 2s.

732. Pair of figures, mounted like fig. 730, namely, in a glass jar, measuring 12 inches by 3 inches, with foot and flanged mouth, with a piece of sheet caoutchouc, 6 inches square, to tie the mouth of the jar when the jar has been *nearly*, but *not quite*, filled with water, and the two imps have been put into it. *Price* 4s.

730.

N.B.—They can be delivered in London, tied over and filled with water; but they cannot be safely sent into the country when filled.

733. The bottle imp is hollow, and has a small opening in one of the feet or in the tail. The weight of the figure is so adjusted that, with the included air, it is a very little lighter than water, and floats at the top of the jar. Nevertheless, when pressure is applied by the hand to the caoutchouc cover of the jar, the image sinks to the bottom of the water. As soon as the pressure is removed the figure rises again to the top. The reason of this is, that the pressure of the hand condenses the air left between the water and the cover of the jar. The condensed air then presses more heavily upon the water, and that in its turn upon the air contained within the image. This air is accordingly condensed into less space, and the image admits a little water, the addition of which makes the image and its contents heavier than the surrounding water, and therefore it sinks. When the hand is removed the compression ceases, whereupon the elasticity of the confined air begins to act. The air within the image, resuming its original volume, expels the water through the hole by which it had entered, and the image, thus restored to its original weight, rises in the water.

If the pressure is continued until the image has too much water forced into it, it will no longer act properly. It should then be wiped dry, and be gradually warmed over a spirit-lamp until the heat forces out of it a sufficient quantity of the superfluous water. Of course, this warming requires a little care, to prevent the cracking of the glass.

Another method of raising imps from the bottom of the jar is to put the jar under a tall receiver on the air-pump, and partially exhaust it; the air within the imps then expands and drives out part of the water, and thus renders the imps lighter and able to float.

EXPERIMENTS ON THE PRESSURE OF THE AIR. 161

When a jar, like fig. 730, is to be set up, the empty imps are to be loaded with water by first gently warming them over a spirit-lamp and then dipping their feet or tail into cold water. Some water then goes into them. If not enough, repeat the process. If too much, expel a little by the method described above. When more water is required, hold the image with its head downwards while you heat it. If water is to be expelled, hold the image with its head upwards while you gently warm it.

Sometimes glass images for this experiment are made in the shape of a balloon. The principle upon which they act is the same, and they require the same management.

"This exceedingly beautiful philosophical toy," says Dr. Arnott, "proves many things: the *materiality* of air, by the pressure of the hand on the top being communicated to the water below through the air in the upper part of the jar; the *compressibility* of air, by what happens in the image just before it descends; the *elastic force* of air shown in expansion, when, on the pressure ceasing, the water is again expelled from the image; the *lightness* of air, in the buoyancy of the image. It shows also that in a fluid *the pressure is in all directions*, because the effects happen in whatever position the jar be held; it shows that *pressure is as the depth*, because less pressure of the hand is required the farther that the image has descended in the water: and it exemplifies many circumstances of *fluid support*. A young person, therefore, familiar with this toy, has learned the leading truths of hydrostatics and pneumatics, and has had much amusement as well as instruction."

734. *Magdeburg Hemispheres*, for showing the pressure of the atmosphere to be about 15 lb. per square inch. The power of the following instruments is found approximately thus:—Multiply the outside diameter by itself, and the product by twelve: the second product expresses pretty nearly the pressure of the atmosphere, or the adhesive force of the hemispheres in pounds avoirdupois; thus

$$3\tfrac{1}{2} \times 3\tfrac{1}{2} = 12\tfrac{1}{4}, \text{ and } 12\tfrac{1}{4} \times 12 = 147 \text{ lb.}$$

Cast-iron hemispheres:—
734 a. $3\tfrac{1}{2}$ inches outside diameter. *Price* 6s.
734 b. $4\tfrac{1}{2}$ inches ditto. *Price* 8s.

Polished brass hemispheres, with brass handles and wooden foot:—
734 c. $2\tfrac{1}{2}$ inches outside diameter. *Price* 10s. 6d.
734 d. $3\tfrac{1}{2}$ inches ditto. *Price* 21s.
734 e. $4\tfrac{1}{2}$ inches ditto. *Price* 35s.

These hemispheres are delivered without stopcocks. When in use, they require a stopcock, with one male and one female screw, fig. 591, the cost of which is 3s.

735. The Magdeburg hemispheres, as represented by fig. 734 consist of two hollow hemispheres of brass or iron, fitting accurately together, furnished with two handles and a stopcock. On removing the moveable handle which is attached to the stopcock, the hemispheres may be screwed to the screw-hole of the air-pump, or to an exhausting syringe, to be exhausted of air by it. The stopcock must then be turned so as to prevent the re-admission of air; and on unscrewing the hemispheres and attaching the handles it will be found that they are firmly held together by the external pressure of the atmosphere, and the united exertions of two very strong men will not effect their separation. It is scarcely prudent to permit boys to tug at the hemispheres; for if they separate suddenly and the boys fall, the hemispheres may be dashed upon the floor so as to damage the edges and put them out of use, especially when the hemispheres are made of cast-iron. But if they are suspended within a receiver, No. 684, by means of the brass ground-plate and hook, No. 624, and the air is exhausted from the receiver, their own weight is sufficient to separate them. A tray or other vessel should be put to receive the falling hemisphere, to prevent its damaging the face of the air-pump table.

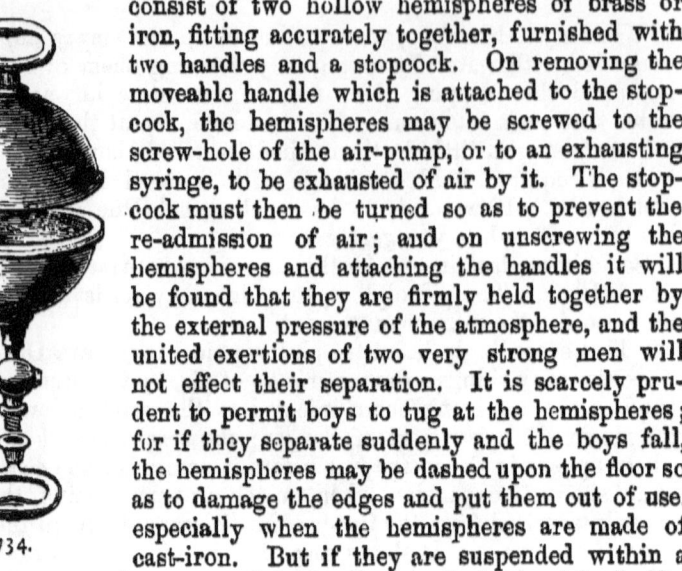

734.

Another proof of the cohesion of the hemispheres is to hang to them a very heavy weight, say 50 lb., which it must be ascertained, before showing the experiment in a Class, the apparatus will carry with certainty.

Before making an experiment, the ground edges of the two hemispheres should be carefully freed from dust, and be greased by means of the tallow holder, No. 638. Of course, pupils must be shown that the hemispheres do not cohere before they are emptied of air.

738. *Glass Model of a Diving Bell*, with chain to lower it into water, and Condensing Syringe, to keep up a supply of air. Price, *without the jar*, 21s.

This apparatus consists of a glass bell, arranged to illustrate the principle of the

738.

common diving-bell. When it is lowered into a jar, filled with water, the action of the condensing syringe serves to supply the bell with fresh air.

739. Deep glass jar for working the above, about 18 inches high and 5 inches wide. *Price* 8s. 6d.

GROUP D.—ON THE MEASUREMENT OF ATMOSPHERIC PRESSURE, § 746 to § 790.

THE BAROMETER.

Prices of fittings for Barometer Experiments, Nos. 746 to 758 :—
746. Support for tubes, fig. 746, black wood. *Price* 4s.
Ditto, mahogany. *Price* 5s. 6d.
747. Glass mortar, or basin, $3\frac{1}{4}$ inch. *Price* 10d.
748. Straight barometer tube, empty. *Price* 1s. 6d.
749. Porcelain cup, with sharp spout. *Price* 1s.
751. Chromo-lithographic scale of 40 inches, with iron stand. *Price* 7s.
751 a. Ditto, scale of 100 centimetres, with stand. *Price* 7s. 6d.
751 b. Syphon or cistern barometer tube. *Price* 2s.
753. Barometer tube and receiver, with caoutchouc stopper, but without mercury basin. *Price* 4s.
754. Barometer tube, with receiver with ground mouth, ground brass plate, and stuffing-box. *Price* 12s. 6d.
755. Bottle, with barometer tube and cork. *Price* 3s. 6d.
756. Tall receiver, 36 inches high, Bohemian glass. *Price* 18s.
758. Bottle, with open tube and caoutchouc stopper. *Price* 3s. 6d.
758 a. Set of six borers for caoutchouc stoppers, with file, in case. *Price* 3s.
781. Slip of whalebone, 3 feet long, for clearing air from mercury in a barometer-tube. *Price* 1s.

746. *The Torricellian Experiment.*—In experiment 721 we have seen the upward pressure of air exerted to sustain water in three glass jars filled quite full and placed in a vertical position, with their mouths downwards.

In fig. 746 we have an experiment represented, in which a column of mercury is sustained by the upward pressure of air in a glass tube which is closed at the top and open at the bottom. The open end of the tube rests in a glass basin containing mercury. The tube is about 33 inches long. The mercury in it rests at a point marked 30″, which signifies 30 inches from the surface of the mercury in the basin. The space above the 30″ in the tube is a vacuum. This figure represents a *temporary barometer*. We proceed to explain the method of preparing it, and how it is that by means of it we are able to prove what is the weight of the atmosphere.

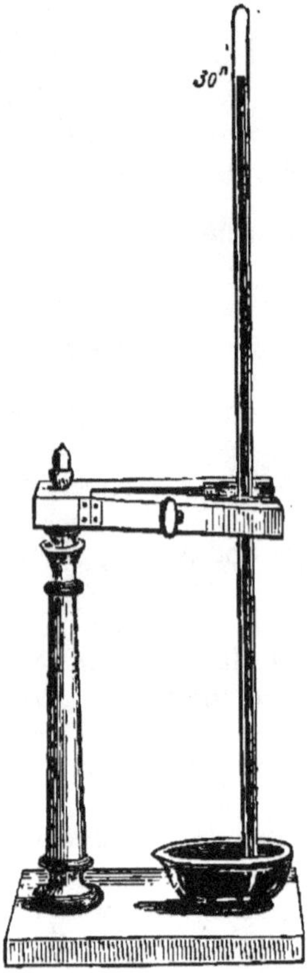

746. 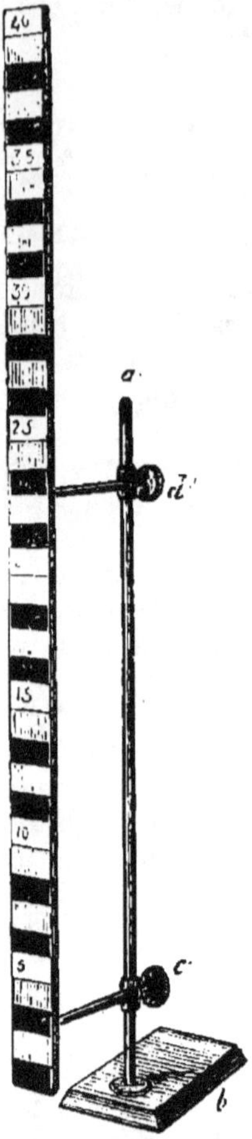   751. 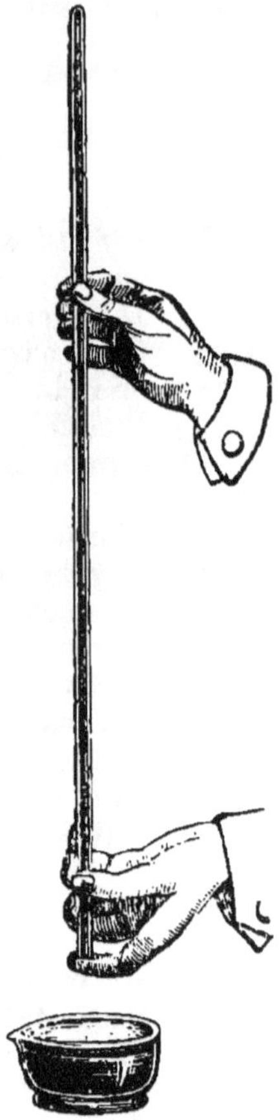   750.

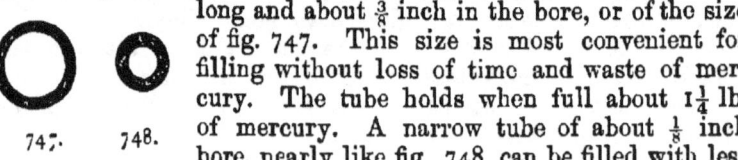

747.   748.

747. *How to fill a Barometer Tube with Mercury.*—Take a clean dry warm glass tube, closed at one end, open at the other, 33 inches long and about $\frac{3}{8}$ inch in the bore, or of the size of fig. 747. This size is most convenient for filling without loss of time and waste of mercury. The tube holds when full about $1\frac{1}{4}$ lb. of mercury. A narrow tube of about $\frac{1}{8}$ inch bore, nearly like fig. 748, can be filled with less

THE BAROMETER.

mercury, but the operation of filling it and also of emptying it is much more troublesome and liable to considerable loss of mercury, in consequence of the difficulty of getting the air out of the tube while the mercury is being passed into it. When it is absolutely necessary to fill a very narrow tube, it can be best done by using a long and very narrow tube as a funnel, the mouth of it being a little widened, in a funnel form. The glass tube being provided, mercury is to be poured into it. The mercury, clean and dry, should be held in a small porcelain cup having a sharp spout, fig. 749; the tube should be held in a slanting position and the mercury be passed into it a little at a time, say as much as will fill 4 inches in length of the tube. The tube should then be held in an upright position and the lower end be gently struck upon a solid piece of felt laid upon the table or the floor; that agitation causes the bubbles of air to collect together and to rise above the mercury. Then another quantity of mercury should be passed into the tube and the air be again shaken out, and that operation is to be repeated until the tube is filled with mercury up to within half an inch of the open end; then you should close the opening with your finger and partially invert the tube so as to bring the sealed end a little higher than the end closed by your finger. The air that was left at the open end of the tube then passes to the sealed end, carrying with it many small bubbles of air that had been retained by the mercury. The level of the tube is then to be altered and the bubble of air to be again passed through the mercury, the tube being meantime revolved a little to bring a fresh surface of mercury upwards. After repeating this operation four or five times, the small bubbles of air will have been carried away. If, however, any remain visible, the process is to be repeated till they are all swept out.

749.

750. Next fill the tube with mercury to the top, brimfull; hold it with your right hand applied near the middle of the tube; close the opening with the fore-finger of the left hand; invert the tube carefully, without letting air into it; plunge the end of the tube below the surface of the mercury contained in a trough, for which a stout glass mortar serves well, and then remove the finger (see fig. 750). Upon the removal of the finger the mercury sinks in the tube to such an extent as to leave in it a column of mercury that measures *about* 30 inches from its upper surface to the surface of the metal in the mortar. The exact number of inches depends upon pressure of the atmosphere at the time of performing this operation, which pressure varies from day to day. Above the 30" of mercury is a *space—a vacuum*. You will observe that there can be no air in that space, for until the finger that closed the orifice of the tube was

taken away in the basin, that space was filled with mercury, and the column of mercury, which was 33 inches high, sank in the tube and left that space of 3 inches free from air; for no air could get into the tube unless it could force its way through the mercury in the basin and through the 30 inches remaining in the tube, or else penetrate through the sealed end of the glass tube. But, since neither of those could be done, it follows that the part of the tube which the mercury leaves must necessarily be a vacuum, unless, indeed, we consider it to be filled with a vapour of mercury, volatilized at mean temperatures.

751. *Scale to measure the height of the mercury in the Barometer*, fig. 751.—This is a chromo-lithographic scale of 40 inches in length, coloured and mounted on a board, as described in § 141. The board is connected with a support, as represented by fig. 751. a is an iron rod, b an iron foot or base, c is a thumb-screw connected with a bar that is attached to the board, and d is a guide intended to keep the scale in a vertical position. The scale can be raised or lowered so as to put the zero of the measure in any desired position. Thus, if it is made to touch the surface of the mercury in the basin or mortar of fig. 746, it will then show the exact height of the mercury in the tube of that apparatus.

751 a. Measures can be supplied, similar to fig. 751, but graduated to 100 centimetres, instead of 40 inches.

751 b. *Syphon Barometer*.—In the syphon barometer the tube and cistern for holding the mercury are in one piece of glass, as represented by fig. 751 b. The *height* of the mercury is the distance between the surfaces at a and at b. The tube should be of the length and width described in § 747, and the cup b of 751 b should be about 1 inch in diameter. The tube is filled with mercury by the following process:—Hold the tube upright, as represented in the figure. Fill the bulb b quite full of mercury, poured into it from the porcelain cup, fig. 749. Close the mouth of b with your thumb, bring the tube into a horizontal position and transfer the mercury, or as much of it as you can, by a series of jerks from the bulb into the tube. Bring back the tube into the position fig. 751 b. Again fill the bulb, and again jerk part of the mercury into the tube. Repeat this process four or five times and the tube will be full. The bubbles of air are to be collected together and swept out by alternately raising and lowering the two ends of the tube, as described in § 747. When the air is all expelled the tube is to be filled as completely as possible, and then raised

751 b.

into the vertical position, fig. 751 b, when the mercury will sink and produce a vacuum at a. The bulb b should be half filled with mercury, and the tube should be put into a support like that of fig. 746. The distance between the two level surfaces of mercury at a and b, measured by the scale, fig. 751, should be 30 inches, or what may happen to be the barometric pressure of the day.

752. *Deduction of the weight of the Atmosphere from the Torricellian experiment.*—The pressure of the atmosphere suffices to sustain a column of mercury of 30 inches in height. If the tube that contains the column of mercury has a horizontal section of 1 square inch, the result is the same as that obtained with the narrow barometer tube. The weight of a column of the atmosphere of 1 square inch in section at the surface of the earth is therefore equal to the weight of a column of mercury of the same section and 30 inches in height. That weight is easily determined thus:— 1 cubic inch of water weighs 252·458 grains, and 1 cubic inch of mercury weighs 13·6 times as much, or 252·458 × 13·6 = 3433·4288 grains, and this, multiplied by 30 for the height in inches, comes to 103,002·8640 grains, or, reduced to pounds, it is 14·7147, or nearly 15 pounds upon the square inch, equal to 2119 pounds on the square foot. The surface of a man's body is commonly assumed to measure 15 square feet. In that case the atmospheric pressure upon him is 2119 × 15 = 31,785 pounds.

753. *Mercury can be raised in a Barometer Tube by atmospheric pressure.*—Fig. 753 represents a barometer tube connected with an air-pump receiver by a perforated caoutchouc stopper. At the bottom of the receiver is a glass mortar, containing rather more mercury than will suffice to fill the tube. The apparatus, as figured, is placed on the plate of the air-pump, and the tube is pulled up so as to make its mouth clear the mercury. The sliding of the glass tube in the caoutchouc stopper takes place easily if the tube is well smeared with tallow. The pump is now to be worked and the air exhausted as completely as possible. That you may know how the exhaustion is proceeding, the pump ought to be provided with a side gauge, such as c, d, fig. 517.

When the exhaustion is complete, there is, or ought to be, no air either in the tube or the receiver. Push the barometer tube gently down into the mercury and slowly admit air into the receiver by the aircock. This air, acting upon the surface of the

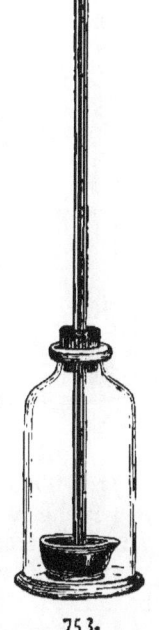

753.

mercury in the basin, drives it up into the barometer tube, higher and higher in proportion to the excellence of the previous exhaustion. Of course, when the barometer No. 746 stands at 30″, the mercury in this apparatus, measured by the graduated scale No. 751, ought also to stand at 30″. But it rarely comes quite so high, because of the imperfect exhaustion both of the receiver and the tube. Nevertheless the experiment shows very clearly that the support of the mercury in the tube is entirely due to the pressure of the air.

754. The apparatus 753 is generally made with brass fittings, including a stuffing-box for the barometer tube. That is a more elegant and more accurate method of mounting it. But the action of the caoutchouc stopper, well greased, is satisfactory.

755. *Mercury is supported in the Barometer by the pressure of the Atmosphere, and it sinks in the tube when that pressure is removed.*—

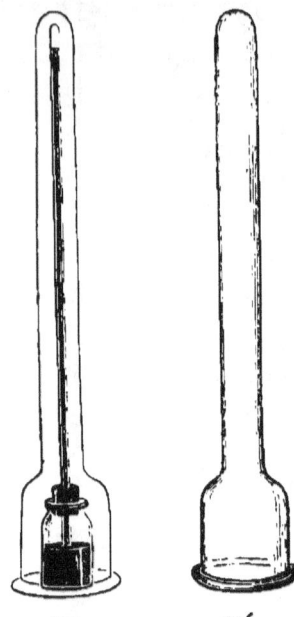

755.   756.

The apparatus required for this experiment is represented by figs. 755, 756. 755 is a narrow, new, and clean barometer tube filled with mercury by the process described in § 747. It is connected with a small wide-mouthed bottle by a cork that holds the tube upright, but does not fit the bottle air-tight. In preparing this apparatus, the tube and bottle are both to be quite filled with mercury, the cork is to be pushed up out of the way, and the tube dipped into the bottle without admitting air—a process requiring a little dexterity. Most of the mercury is then to be poured out of the bottle, leaving only enough to cover the mouth of the tube. There must be enough space left in the bottle to receive all the mercury contained in the tube when it is subsequently made to descend. Adjust the cork so as to keep the tube upright, put the bottle on the plate of the air-pump, and cover it with the large receiver, fig. 756.

756. This receiver measures about 36 inches high, the tube of it is 2 inches wide and about 30 inches long; the lower part measures 5 inches across the ground flanges.

757. The apparatus being properly adjusted, the air-pump is to be worked. The mercury in the tube immediately begins to descend, and as the exhaustion continues, the mercury constantly descends

until it has all left the tube and gone into the jar. There will then be a vacuum in the receiver. A column of 30 inches of mercury was suspended in the barometer tube when the air pressed freely on the mercury in the bottle, but the mercury sank as the pressing air was withdrawn, and none was left in the tube when the receiver was fully exhausted.

The conclusion, that it is the weight of the air that supports the mercurial column is confirmed by the admission of air into the receiver by opening the aircock of the pump, for the air then rushes into the exhausted bottle, and pressing the mercury into the tube, makes it rise till it stands therein at its former height of 30 inches.

758. *The elasticity or spring of the Air is equal to its compressing force.*—The apparatus required for this experiment is represented by figs. 755, 756, excepting that 755 is mounted as follows:—The bottle is to be half filled with mercury, the tube to be open at both ends, and dip deeply into the mercury; and the stopper must be of caoutchouc, and fitted air-tight both to the bottle and the tube.

It may not be amiss to state here that perfectly round holes can be cut in caoutchouc stoppers by using brass cork borers, fig. 758 a, filed pretty sharp on the cutting-edges a and thoroughly wetted with spirit of wine. In use, the joints between the glass tubes and the caoutchouc should be greased with tallow.

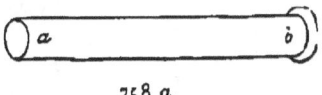

758 a.

Process.—Place the apparatus, fig. 755, corrected as above described, on the table of the air-pump, cover it with the receiver, fig. 756, and proceed to exhaust the air from the receiver. As that exhaustion goes on, the air in the bottle expands, and pressing upon the surface of the mercury, forces it up the tube. When the exhaustion is completed, the mercury will be nearly at barometer height in the tube. On admitting air into the receiver, the mercury descends from the tube into the bottle.

MARIOTTE'S APPARATUS FOR SHOWING THE COMPRESSIBILITY AND ELASTICITY OF AIR.

780. *The volume of a given weight of air is inversely as the pressure to which it is exposed;* that is to say, the greater the pressure to which the air is subjected the smaller becomes its volume, and the less the pressure the larger is the volume. This is commonly called the *law of Mariotte.* It was discovered independently by Mariotte in France, and Boyle in England, about the year 1670.

We demonstrate the truth of this law by two experiments: one, fig. 781, to illustrate the effects of pressure *greater* than that of the atmosphere; the other, fig. 782, to illustrate the effects of pressure *less* than that of the atmosphere.

781. *Mariotte's Apparatus, constructed to show that, under the pressure of two atmospheres, air is compressed into half its ordinary bulk,* fig. 781. *Price* 18s.

This apparatus consists of two graduated glass tubes connected by iron fittings. The short tube is graduated to show its cubical contents, and has twenty divisions in two spaces of ten each. The long branch is graduated to show 32 linear inches. The tubes are both about $\frac{1}{2}$ inch in diameter. The iron base is perforated, so that the two glass tubes are in direct communication with one another. The apparatus requires $2\frac{1}{2}$ lb. of mercury to work it.

Use.—By means of the key, k, unscrew the tubes, and see that the leather washers are in good condition. If not, replace them with new ones (see § 611). Then screw up the tubes, shut the stopcock, unscrew the cap on the short tube, and, with the help of a small funnel, put into the apparatus as much mercury as will rise up to 20 in the short tube and to a in the long tube. Screw on the cap of the short tube. You have now, confined in that short tube, 20 measures of air, acted upon by the pressure of the atmosphere at a. Pour into the long tube as much mercury as will rise to the mark 30 inches near the top of the scale. Remove the air bubbles by dipping into the mercury a long narrow slip of whalebone (not a metal wire), and adjust the mercury at 30 inches. The air in the short tube will then be level with 10 on the scale; consequently 20 parts of air, when under the pressure of one atmosphere, become 10 parts when under the pressure of two atmospheres.

781.

781 *a*. The apparatus, fig. 781, is an example of the communicating vessels with two liquids, which were described at § 215. In the first part of the above experiment we have a level line of junction seen at 20 in one tube, and a in the other, and the liquids in equilibrium are the uncondensed air in the short tube, and the atmosphere in the long tube. In the second part of the experiment we have the level line of junction

MARIOTTE'S APPARATUS.

elevated to 10 in the short tube and to 0 in the long tube, while the liquids in equilibrium are the 10 volumes of condensed air in the short tube, and in the long tube a compound of 30 inches of mercury *plus* the pressure of the atmosphere. We see, therefore, that the 10 measures of condensed air have a power of pressure or resistance equal to that of two atmospheres, or to 60 inches of mercury.

When the experiment with fig. 781 is concluded, the mercury should be removed from it. The base of the apparatus should be placed near the edge of a table with the stopcock projecting beyond it, and a glass bottle be so held below the stopcock, that when the latter is gently opened the mercury may be collected in the bottle without loss.

781 *a*. Price of slip of whalebone, 3 feet long, 1*s*.

782. *Mariotte's Apparatus, constructed to show that, under the pressure of half an atmosphere, air expands to twice its ordinary volume,* fig. 782. *Price* 12*s*.

This apparatus consists of two glass tubes, *a* being about 22 inches long by ¼ inch wide, and *b* about 24 inches long by ¾ inch wide. *a* is mounted with a small stopcock, and is graduated, first, into two equal cubical spaces, marked 1 and 2; and secondly, with a scale of 16 linear inches, beginning with 0 at the mark 2. *b* is supported by a large and heavy wooden foot, about 6 inches wide and 5 inches high. This apparatus, like fig. 781, requires 2½ lb. of mercury to work it.

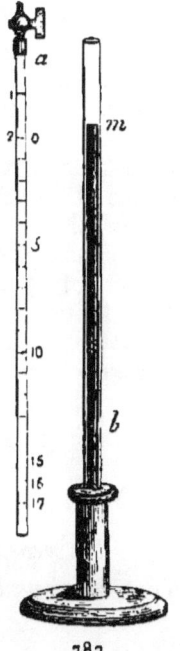

782.

Use.—Fill the tube *b* with mercury up to about 2 inches from the top. Remove air-bubbles by means of the slip of whalebone mentioned in the preceding article. Then open the stopcock of the tube *a*, and slowly plunge the graduated open end of that tube into the mercury contained in the tube *b*. It must go down so far as to cause the mark 1 to be on a level with the surface of the mercury then present in both tubes. To make this convenient, the mercury in the tube *b* should come to about an inch from the top when the tube *a* is immersed in it. When this point is adjusted, the stopcock of the tube *a* is to be closed. You have then in that tube 1 measure of air at atmospheric pressure. Pull up tube *a* slowly with the mercury which rises within it until mark 2 is found level with the surface of the mercury contained in tube *a*. Then, upon reading the scale of inches engraved on tube *a*, commencing at 0 at the level of 2 on the same tube, and descending, you will find that the surface of the mercury in the outer tube rests where 15 inches is marked on the inner tube; consequently the volume of the air enclosed in tube *a* is enlarged from *one* to *two* volumes, while the atmospheric pressure is reduced to 15 inches of mercury.

As the figures on the narrow tube *a* are necessarily very small, it is convenient to show the space of 15 inches between the two mercury levels in *a* and *b*, by using a chromo-lithographic scale of 20 inches, mounted on wood, as described at § 141.

782 *a*. We have shown in § 781 that the apparatus there described may be taken as an example of the communicating vessels with two liquids which were described at § 215. The same may be said of the apparatus represented by fig. 782.

In the first part of the experiment the tube *a* is plunged into the mercury contained in tube *b* (the stopcock of *a* being first opened) until the mercury in both tubes is level at the mark 1 on the small tube, that tube being held steadily in its place to prevent its swimming. At that moment the level of the mercury in both tubes is the horizontal line of junction, and the atmospheric air pressing upon the mercury in both tubes is in equilibrium.

In the second part of the experiment the stopcock *a* is closed and the small tube is lifted up with the mercury that rises with it until there is a difference of 15 inches in the two levels of the mercury. At that moment the horizontal line of junction is level with the surface of the mercury in the tube *b*, and with the line of 15 inches in the tube *a*. Above that line of junction there rests in tube *b* the pressure of one atmosphere, and in the tube *a* there is first the pressure of 15 inches of mercury, equal to half an atmosphere, and above that the pressure of the expanded air, which must be equal to another half atmosphere, in order to complete the pressure necessary to produce equilibrium.

GLASS RECEIVERS FOR AIR-PUMPS,

All with Flanged Rims, Ground Flat, ready for use.

The measurement of Width includes the Ground Flanges.

The measurement of Height is from the base to the centre of the Dome, inside the Receiver.

796. The flat ground flange of the receiver that rests upon the plate of the air-pump commonly measures from $\frac{1}{3}$ inch to $\frac{1}{2}$ inch across. The bore of every receiver is therefore from $\frac{2}{3}$ inch to 1 inch less than its extreme width over the flange or welt. It is

GLASS RECEIVERS FOR AIR-PUMPS.

this extreme width over the flange that serves to show for what air-pump plate the receiver is suitable. Thus a 7-inch air-pump plate (see No. 505) will take a receiver that has an extreme width of 7 inch, or any smaller size; but it will not work with any receiver that is wider than 7 inches, not even with one of $7\frac{1}{4}$ inches.

In selecting receivers for the air-pump, the *width* is therefore the main thing to which attention must be paid. The *height* may be variable, and indeed is necessarily variable for different apparatus. As a general rule the receiver should be so small as merely to cover the object submitted to experiment, because the smaller it is the less work there is for the pump to do in exhausting the air from it. This is sometimes important, as, for example, in the experiment of the freezing of water, § 700, in which case a small flat receiver is essential. On the contrary, for the trial of the bell in vacuo (see §§ 695, 696), a large receiver best answers the purposes of the experiment. When, again, an experiment is to be made at a lecture, and to be seen by a large audience, it is proper to have some space between the apparatus set in action and the dome of the receiver that covers it, otherwise the apparatus is liable to be rendered invisible by the play of light from the glass dome. Finally, there is the fact to be dealt with, that every operator wishes to have as few receivers as possible, in order to limit his expenses. In the following list some care has been taken to point out the most useful forms and sizes, and the operator who compares this list with the figures and descriptions in the chapter on Pneumatic Experiments, will be readily led to select what are most suitable for his purposes.

797. All the receivers described in this list are of *hard glass*, not flint glass. Those in the columns headed B are of fine Bohemian glass, and handsomely finished. Those in the column headed G are of hard German glass, not quite so clear from specks and free from colour as the Bohemian glass. The whole are well annealed.

798. Exact sizes are specified in the lists of glass vessels in order to guide purchasers in their choice, and when instructions are received by the advertisers to supply vessels of special sizes, these instructions will be attended to as closely as possible; but at the same time no guarantee can be given that in such cases the measurements shall be correct to small fractions of an inch, for it is impossible to have glass vessels manufactured with such scrupulous exactness in non-essential particulars.

Very great changes in *the price of glass* have taken place during the years 1872-73. While this sheet was in the press, a rise of 15 per cent. occurred close upon previous rises; consequently the prices in this list must be considered as *the prices of the day*, but not unalterably fixed.

PNEUMATICS.

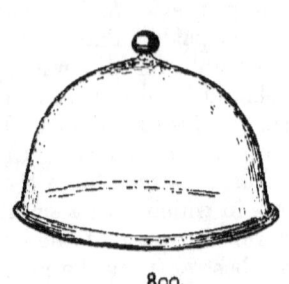

800.

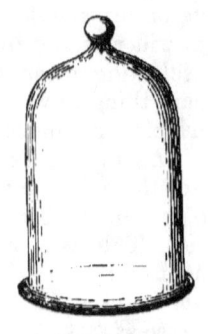

809.

Flat Receivers for Air-pumps.   *Cylindrical Receivers.*

No.	Inches wide.	Inches high.	PRICES. B.		PRICES. G.		No.	Inches wide.	Inches high.	PRICES. B.		PRICES. G.	
			s.	d.	s.	d.				s.	d.	s.	
800	6½	3½	4	0	2	6	809	6½	8	5	0	3	
801	7	4	4	6	2	6	810	7	8	6	0	3	
802	8	6	7	0	3	6	811	7	12	9	0	5	
803	10	6	11	6	4	6	812	8	12	10	6	6	
804	11½	7	17	0	6	6	813	10	14	22	0	..	
							814	4	6	2	6	1	
							815	6	20	25	0	..	
							816	5	36	38	0	..	

GLASS RECEIVERS FOR AIR-PUMPS.

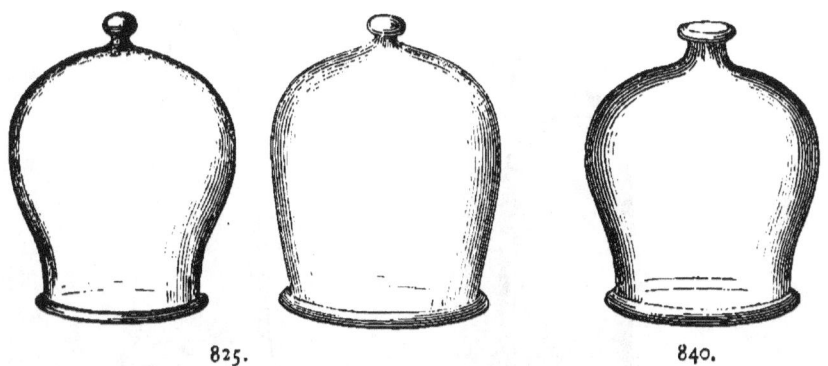

825.
840.

Bell-shaped Receivers. *Bell-shaped Receivers with necks.*

No.	Inches wide.	Inches high.	PRICES. B.		PRICES. G.		No.	Inches wide.	Inches high.	PRICES. B.		PRICES. G.	
			s.	d.	s.	d.				s.	d.	s.	d.
825	4	6	3	0	2	6	840	4	6	3	6	2	6
826	6½	8	7	6	3	6	841	6½	8	8	0	3	6
827	7	8	9	0	4	0	842	7	8	10	0	4	0
828	8	10½	14	0	6	6	843	7	12	13	0	6	6
829	10	12	24	0	9	0	844	8	10½	15	0	6	6
							845	10	12	25	0	9	0

The necks of the receivers, fig. 840, are 3 inches in bore, and 4 inches across the ground flanges. They are adapted to the brass plate of the hook, fig. 624, and the sliding-rod, fig. 625.

176 PNEUMATICS.

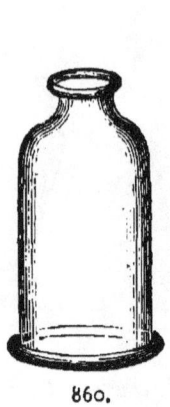

860.   880.

Cylindrical Receivers with necks.   *Conical Fountain Glasses.*

No.	Inches wide.	Inches high.	PRICES. B. s. d.	PRICES. G. s. d.	No.	Inches wide.	Inches high.	PRICES. B. s. d.	s.
860	5	14	9 0	4 0	880	4	15	7 6	4
861	7	12	12 6	5 0	881	5	15	12 6	5
862	10	14	24 0	..	882	5	21	18 0	6

The necks of the above are 3 inches in bore, and 4 inches over the ground flanges. Like fig. 840, they fit the brass plate of the hook, No. 624, and the sliding-rod, No. 625.

| 866 | 4 | 10 | 5 6 | 2 6 |

With neck of 2½ inches bore, and 3½ inches over the ground flanges. Used for the filter cup, No. 712.

GLASS RECEIVERS FOR AIR-PUMPS.

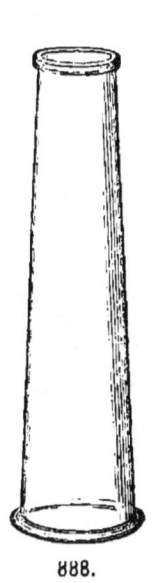

888.

Guinea and Feather Glasses.

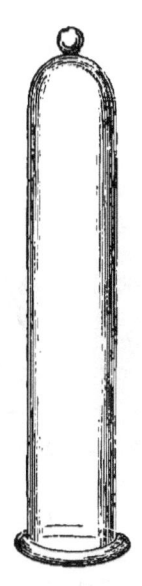

815.

815. *Tall Cylin*
895. *Tall plain*

Wide at bottom.	Wide at top.	Inches high	PRICES.				No.	Inches wide.	Inche high
			B.		G.				
			s.	d.	s.	d.			
$4\frac{1}{2}$	4	16	15	0	5	0	895	4	3c
5	4	18	18	0	6	0	896	4	16
5	4	24	21	0	7	6			

MISCELLANEOUS GLASS FITTINGS,

For Pneumatic and Hydrostatic Experiments.

901. Bohemian glass basins, flat bottoms, with vertical sides, fig. 901.

No.	Wide.	Deep.	Contents.	Price.
901.	6 inches	$3\frac{1}{4}$ inches	2 pints	1s. 3d.
902.	8 ,,	5 ,,	6 ,,	2s. 6d.
903.	9 ,,	$5\frac{1}{2}$,,	8 ,,	3s. 3d.
904.	10 ,,	6 ,,	10 ,,	4s. 6d.
905.	15 ,,	7 ,,	28 ,,	15s.

913. Deep conical stoneware pans, salt glazed, fig. 913.

913.	Diameter 8 inches, contents 3 pints.	Price 1s. 8d.
914.	,, 10 ,, ,, 5 ,,	,, 2s.
915.	,, 12 ,, ,, 8 ,,	,, 2s. 6d.

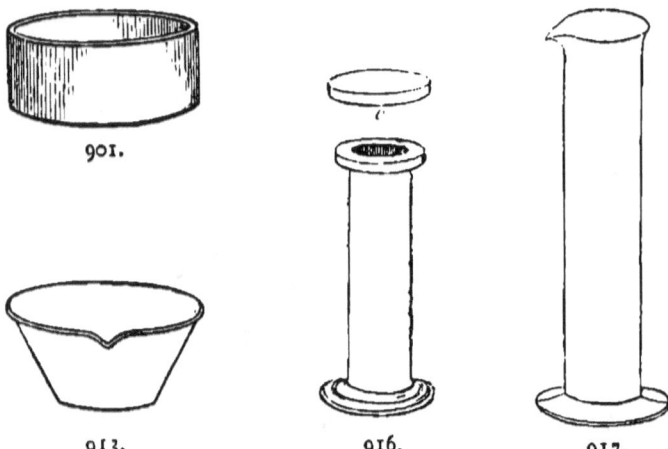

901.   913.   916.   917.

916. Glass jars on feet, hard German glass, with flanged mouth, fig. 916, or with spout, fig. 917, *at the same price.*

Sizes as follow:—

No.	High.	Wide.	Price.
916.	6 inches	2 inches	10d.
917.	8 ,,	2 ,,	1s.
918.	10 ,,	2 ,,	1s.

MISCELLANEOUS GLASS FITTINGS.

Glass jars on feet, *continued.*

No.	High.	Wide.	Price.
919.	12 inches	2 inches	1s. 4d.
920.	15 ,,	2 ,,	1s. 8d.
921.	8 ,,	2½ ,,	1s. 4d.
922.	10 ,,	3 ,,	1s. 6d.
923.	12 ,,	3 ,,	1s. 9d.
924.	10 ,,	4 ,,	2s. 6d.
925.	16 ,,	5 ,,	6s.

Many other sizes can be supplied.

916 c. Circular disks of stout plate glass, polished on both sides and on the edges, fig. 916 c:—
3 inch, *price* 1s. 4 inch, 1s. 6d. 5 inch, 2s. 6 inch, 3s.

931. Wide-mouthed bottle, to use as water reservoir in fountain experiments, size 12 inches high, 5 inches wide, mouth 2½ inches in bore. See A, fig. 516. *Price* 3s.

936. Bohemian glass beakers for holding liquids, either hot or cold, large set of 15 beakers, holding from 1 ounce to 220 ounces, fig. 936. *Price* 16s.

937. Ditto set of 8, from 8 to 100 ounces. *Price* 9s.

938. Ditto set of 5, from 3 to 18 ounces. *Price* 3s.

941. Griffin's beaked tumblers, with spouts, set of 6, contents from 5 ounces to 40 ounces, fig. 941. *Price* 5s.

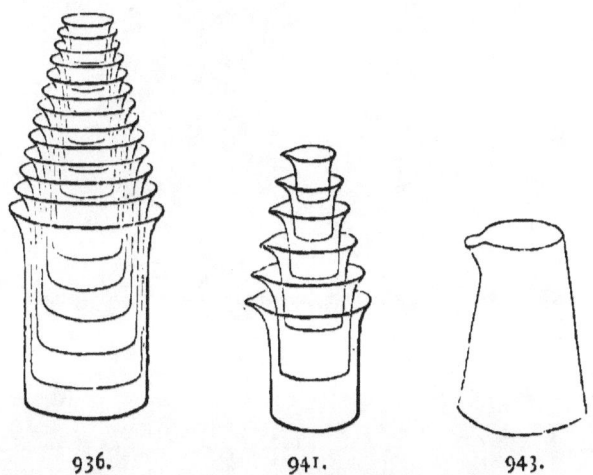

936.   941.   943.

943. Conical beakers, with spout, fig. 943 :—

| ¼ pint, *price* 8d. | 1½ pint, *price* 1s. 3d. | 4½ pints, *price* 2s. 6d. |
| ½ pint, *price* 10d. | 3 pints, *price* 1s. 9d. | 9 pints, *price* 3s. 6d. |

945. Glass funnels, fig. 945:—
   1 inch, *price* 1½*d*.   4 inch, *price* 4*d*.
   2 inch, *price* 2*d*.   6 inch, *price* 10*d*.
946. Funnel holders for small funnels, black wood. *Price* 1*s*.
947. Ditto for large funnels, mahogany. *Price* 5*s*.

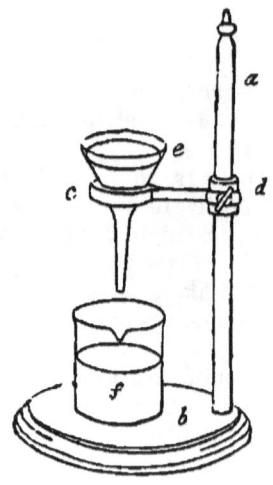

INDEX.

ADHESION plate, 19
Air, difficulty of extricating it from water, 142; and from mercury, 165
 its elasticity equal to its compressing force, 169
 its extrication from various bodies, 146
 weighed, 136
Air-pumps, 111
 Tate's, mounted on clamp, 115
  on pedestal, 121
  on table, 122
  large size, 123
   with fly-wheel, 124
   with extra fittings, 122
  compared with other pumps, 119
  details of construction, 116
  with two vertical barrels, 114
  with two barrels, raised plate, 115
  with three barrels, 124
  one barrel, small size, 126
  exhausting power of, 112
  extra fittings for, 128
  instructions for care of, 127
  labour of working different kinds, 126
Air syringes, brass, 127
 condensing, 128
 exhausting, 127
 exhausting and condensing, 128
  with clamp, 128
Alcoholometer, 75
 Gay-Lussac's, 78
 Sikes's, 79
 Tralles's, 78
Ale, air from, 147
Apple cutter, 148
Appold's centrifugal pump, 109
Archimedes, principle of, 57
 experimental demonstration of, 59
Archimedian screw, glass, 109
 model in metal, 109

Atmosphere, weight of, iron bar to represent, 157
 weight of, how deduced from the Torricellian experiment, 167
Atmospheric optic marvel, 26
 pressure, crushing power of, 157
  measurement of, 163
Attwood's fall machine, 20
Aurora borealis apparatus, 135

Balance extemporised, 3
 hydrostatic, 32
  Mohr's, 69
 for accurate weighings in case, 33
 in mahogany box, 34
 commercial, 34
Balloons, 139
Barker's mill, in glass, 48
 in metal, 47
Barometer, 163
 to fill with mercury, 164
 support for, 164
 scale for, 166
 syphon, 166
 mercury raised in it by atmospheric pressure, 167
 effect of removing that pressure, 168
Baroscope, 140
Batavian tears, 31
Beaker glasses, 179
Beer, air from, 147
Bell shaped receivers for air-pumps, 175
 with necks, 175
Bells to be rung in vacuo, 150
Black rail, 1
Bladder experiments, 148
 frame and lead weights, 149
 glass, 155
 experiments with, 156
 piece, 132
Bladders, preparation and care of, 149

Blank nut, 131
Bohemian beakers, 179
Bolognian flasks, 31
Bottle imps, 58, 160
Bottles, specific gravity, 65
Bottle to hold $\frac{1}{100}$ gallon, 64
Bowl-about, 15
Boyle's inverted U tube, 43
Bramah press, glass, 50
  small brass, 50
  large brass, 51
Brass pulleys, set of, 10
Breaking squares, 150
Breast water-wheel, 109
Brittleness, 31
Bulb gauge and jar, 144
Bulb pipettes, 86
Burette, 86
  Binks's, 86
  Gay-Lussac's, 86
  Mohr's, 86

Caoutchouc balls, 134, 149
Capillarity, 81
Capillary plates, 82
  tubes, 82
Caps for air jars, 132
Carbonic acid gas weighed, 139
Cartesian devils, 160
Cathetometer, 31
Central forces, laws of, 22
Centre of gravity, 14
  instruments for showing, 14
    bowl-about, 15
    brass semicircle, 14
    double cone, 15
    equilibriste, 15
    irregular board, 14
    leaning tower, 14
    parallelopipeds, 14
Centrifugal machine, 22
  adjuncts for 12 experiments, 23
Centrifugal pump, Appold's, 109
  railway, 29
Chromatic fire-cloud, 102
Chromo-lithographic scales, 36, 166
Clamps, 135
Clinometer, 32
Clip and weight, 133
Coke, air extricated from, 147
Collision, 19
Communicating vessels, 37
  for one liquid, 37
  for two liquids, 40, 41
  for three liquids, 42

Compound wheel and axle, 12
Condensed air fountains, 99
  with glass reservoir, 99
    large zinc reservoir, 99
    small brass reservoir, 101
  jets for, 101
  syringes to supply air, 102
Cone for illustrating stability and instability, 16
Cone of water weighed, 55
Conical beakers, 179
Conical fountain glasses, 176
Connectors, brass, 130
  1, 2, 3, and 4 way, 131
Constructive mechanics, 29
Cord for pulleys, 2
Cork, air extricated from, 147
Cork and balance weight, 140
Crimson dye for water, 38
Crushing weight of the atmosphere, 157
Cylindrical receivers for air-pumps, 174
  with necks, 176

Dialyser, Graham's, 31
Diving-bell, 162
Double cone, to roll up hill, 15
Double transferer, 134
Downward pressure of the air, 154

Elastic balls, 149
Equilibrium of a floating body, 80
  of solids, 16
  of solids when immersed in liquids, 57
Equilibriste, 15
Evaporation in vacuo, 153
Exhausting powers of air-pumps, 111
Expansion of air, 144
Eye fountain, 93

Falling bodies, laws of, 20
Filter cup for mercury shower, 154
  for shower of air, 155
Filtering-paper, 82
Filtering-tube to show capillarity, 82
Fire engine, glass model, 95
  metal, 95
Fire syringe, 159
Flat receivers for air-pumps, 174
Floating bodies, weight of, estimated by measuring the water they displace, 79
Flotation, 79

Force pumps, 94, 95, 96
Forge hammer, 30
Fountains, 97
  produced by a fall of water from an eminence, under common atmospheric pressure, 98
  when condensed air forces a water jet into free air, 99
  when uncondensed air forces a water jet into an exhausted receiver, 102
  when a column of water condenses a confined portion of air and causes it to force a water jet into free air, 104
  when a fall of water from a syphon creates a vacuum into which a water jet is forced by free air, 105
  when an interrupted supply of air causes a spring to be intermittent, 107
Fountain glasses, 176
  syphon, 93
Freezing of water by the air-pump, 152
Fruit stand, 148
Fulcrum for lever, 2
Funnels, 180

Glass basins, 178
Glass breaking squares, 150
Glass jars on feet, 178
Governor, Watts's, 25
Graduated measures for ounces, 35
  for grammes, 35
Graham's dialyser, 31
  osmometer, 31
Gravesende's lever, 6
Gravity, centre of, 14
Grease for air-pump receivers, management of, 136
Ground brass plate and hook, 133
  with sliding rod, 133
Guinea and feather experiment, 142
  brass work for, 142
  glasses for, 142, 177
  long tube for, 143
Gyroscope, 27
  small size, 28
Gyroscopic top, 28

Haldat's hydrostatic apparatus, 54
Hand glass, 155
Hempel's syphon, 92

Heron's ball, 103
  fountain, 104
  glass model, 104
  metal model, 105
Hooke's universal joint, 30
Hooks for blackboard, 1
Houdin's inexhaustible bottle, 88
Hydraulics, 81
Hydrodynamics, 81
Hydrogen gas weighed, 139
Hydrometers, 73
  forms of, 73, 74, 76
  jars for, 73, 74, 76
  Baumé's, 77
  Nicholson's, 72
  Twaddell's, 77
  special scales for, 77
  rules for using, 78
  price list of, 78
Hydrostatic balance, 22
  Mohr's, 69
  bellows, 49, 50
  paradox, 52, 57
  press, 50
  principle, Pascal's, 53
Hydrostatics, 37

Inclined plane, 10
Indigo blue dye for water, 38
Inertia illustrated, 27
  apparatus, 16
Intermittent fountain, 107
  motion, 29
  spring, 107
Iron pins and hooks, 1
Iron connectors, 130
Iron stopcocks, 130
Irregular board to show centre of gravity, 14

Key, 135, 170

Lactometer, 79
Lateral pressure of liquids, 47
Laws of falling bodies, 20
Leaning tower, 14
Leslie's apparatus for freezing water, 152
Level of different liquids in communicating vessels, 39
  of junction of two liquids, 39
  of water in communicating vessels, 37, 38, 39
Lever, 2
  of the first kind, 4

Lever:
  of the second kind, 5
  of the third kind, 6
  Gravesende's, 6
  for suspension, 3
  with solid fulcrum, 2
  varieties of, 4
Lift pump, 94, 95, 96
Linear measures, inches, 36
  centimetres, 36
Lock and key, 31
Luugs glass, 146

Magdeburg hemispheres, 161
Magic bottle, 158
  can, 88
Mariotte's apparatus, for showing the compressibility and elasticity of air, 167
  constructed to show that, under the pressure of two hemispheres, air is compressed into half its ordinary bulk, 170
  and that under the pressure of half an atmosphere it expands to twice its ordinary volume, 171
Mechanical powers, 1
  small set in box, 13
Mechanics, 1
Mercury, shower of, 155
Mohr's hydrostatic balance, 69
Mysterious funnel, 87

Nicholson's hydrometer, 72

Oblique cylinder, 14
Optic marvel, 26
Osmometer, 31
Overshot water-wheel, 108

Pair of narrow bottles for transfer of air, 145
Pan and peas, 16
Pans, 178
Parallel motion, 29
Parallelogram of forces, 18
Parallelopipeds, 14
Pascal's hydrostatic apparatus, 53
Pendulum, 16
Percussion machine, 19
Phial of the four elements, 43
Philosophical hammer, 141
Pipettes, 83
  bulb, 86
  scale, 86

Pipettes:
  various forms of, 84
  safe use of, 84
  prices of, 85
Pipette toys, 87
Plants, air extricated from, 147
Plumb-line, 16
Pneumatics, 111
Porous evaporating-basins, 152
Pressure of fluids proportionate to depth, 83
Pressure of the air, 154
Pressure of water in all directions, 44, 45, 47
Prince Rupert's drops, 31
Pulley, 7
  simple, 7
  fixed, 7
  moveable, 7
Pulley frame, 7, 10
Pulleys, set of, wood, 7
  set of, brass, 10
  system of, 8
  concentric, 10
  long three-sheave, 8
  square three-sheave, 10
Pumps, Water-, 94
  cause of the rise of water in, 47, 97
  lift-pump, glass, 94
  brass mounts, 96
  Tate's school pattern, 97
  force-pump, glass, 94
  larger, 95.
  double action, glass, 95
  metal, 95
  brass mounted, 96
  mahogany stand for, 97

Receivers for air-pumps, 172
Regnault's specific gravity bottle, 65
Resistance of air, 156
Rotating apparatus, 27
Rupert's drops, 31

Saccharometer, 79
Scale pans, 2
Scale pipettes, 86
Schoolroom blackboard, 1
Screws 13
Semicircle of brass, to show centre of gravity, 14
Shower of air, 155
Simple pulley, 7
Single transfer plate, 133
  in use, 102

INDEX.

Sliding rod in stuffing-box, 133
Smoke jacks, 150
Specific gravity, principle of, 57
  experimental illustrations, 59
  difference in liquids, 58
    in solids, 58
  of solids and liquids, method of estimating, 61
  of solids, determined by the hydrostatic balance, 61
  of liquids, determined by the hydrostatic balance, 62
  of a solid, determined by weighing it in a bottle, 62
  of a liquid determined by weighing it in a bottle, 63
  bottles for solid bodies, 62
    for liquids, 63, 65
  bottles, on the choice and use of, 67
  various forms of, 65, 66.
Specific gravity of a liquid estimated by weighing in a balance a quantity first measured by a pipette, 69
Specific gravity of a solid estimated by measuring the water it displaces, 71
Specific gravity of a liquid estimated by means of the hydrometer, 73
Specific gravity of gases, 136
Specific gravity, table to facilitate calculations, 68
Spirit level, 39
Spouting jars for illustrating Torricelli's theorem, 83
Springs and fountains, classification of, 97.
Stability and instability of floating bodies, 80
Stopcocks, pneumatic, 130
Suspension board, 1
Syphon, 89
  various forms of, 89
  principle of action, 90
  use in fitting complex forms of chemical apparatus, 91
Syphon barometer, 166
Syphon fountain, 105
Syphon gauges, powers of, 112
  various forms of, 129
Syphon toys, 92
Syringes for air, 127, 128
  for water, 93
Syringe and lead weight, 158
Systems of pulleys, 7, 8, 10

Tall cylindrical receivers, 177
Tall plain jars, 177
Tallow holder, 135
Tantalus's cup, 92
Taper stand, 148
Three-globe apparatus, 159
Tilt hammer, 30
Top of tops, 28
Torricellian experiment, 163
Torricelli's theorem, 83
Tourniquet hydraulique, 48
Toys founded on pipettes, 87, 158
  founded on syphons, 92
  to illustrate adhesion, 19
Train of wheels, 29
Transfer plates, single, 133
  double, 134

Undershot water-wheel, 108
Universal joint, 30
Upward pressure of the air, 157
Upward pressure of water, 48, 49

Valves, 95
  butterfly valve, 96
  bellows valve, 96
  round spring valve, 96
  conical valve, 96
  oil silk valve, 96
Velocity of falling bodies, 20
Vernier, 32
Vibrating wire, 28
Volumetric analysis, 83

Washers, 132
Water, difficulty of freeing it from air, 142
  falls solid in vacuo, 141
  freezing it in vacuo, 152
  pressure in all directions, 44
  lateral pressure, 47
  upward pressure, 48, 49
  pressure proportionate to depth, 83
  pressure on the two ends of a cone, 57
  incompressibility, 46
  weighed in a vessel with a loose bottom, 55
  rises to a level in communicating vessels, 137
  to colour blue, 38
  to colour red, 38
Water hammer, straight, 141
  V-formed, 141
  fork-formed, 142

INDEX.

Water-pumps, 94
Water-wheels, 108
Waterworks, 108
Wedge, 11
Weights, 34
  for levers and pulleys, 2
  grain weights, accurate, 34
  grain weights, not adjusted, 34
  pound pile, brass, 34
    cast-iron, 34
  Centigrade, 35

Wheel and axle, 11
  compound, 12
Whirling table, 22
  adjuncts for twelve experiments, 23
White's concentric pulley, 10
Windmill, 143
Wire cage for breaking squares, 150
Wire gauze capsule, 82
Wooden balls to show capillary action, 82
Würtemberg syphon, 91, 93

THE END.

SCIENTIFIC BOOKS AND CATALOGUES

PUBLISHED BY

JOHN J. GRIFFIN & SONS,

22, *GARRICK STREET, LONDON*, W.C.

CATALOGUE OF SCIENTIFIC APPARATUS for USE in SCHOOLS:
Comprehending Instruments required for Performing Experiments to illustrate the following Sciences: Mechanics, Hydrostatics, Hydraulics, Pneumatics, Acoustics, Meteorology, Heat, Light, Electricity, Magnetism, Galvanism, Electro-magnetism, Chemistry, Geology, Crystallography, Mathematics. 56 pages, demy 8vo. Price 6d. post free.

STUDIES IN CHEMICAL PHILOSOPHY,
Embracing Discussions respecting the Constitutions of Acids, Bases, and Salts; the Atomic Weights, the Radical Theory, the Constitution of Gases, the Construction of Chemical Formulæ, the Principles of Chemical Nomenclature, the Doctrine of Types and Substitutions, Special Account of the Constitution of most of the important Varieties of Salts, &c. &c. &c. By JOHN J. GRIFFIN, F.C.S. Crown 8vo., 558 pages. Price 5s., cloth.

ECONOMY IN SUGAR MAKING.
Directions for Testing Cane Juice, so as to determine the exact quantity of Quicklime required to temper the juice. By the late JOHN SHIER, LL.D., Agricultural Chemist to the Colony of British Guiana. A new edition, edited by JOHN J. GRIFFIN, F.C.S.
[*In the Press.*
The Apparatus for putting the testing into operation is supplied by the Publishers of this Work.

CHEMICAL LABELS:
1. Set of 600 Labels, small type, 6d. 2. Book of 500 Labels, large type, 1s. 6d.

SCIENTIFIC WORKS by JOHN J. GRIFFIN, F.C.S.

CHEMICAL RECREATIONS:
A POPULAR
MANUAL OF EXPERIMENTAL CHEMISTRY.

By JOHN JOSEPH GRIFFIN, F.C.S.

The TENTH EDITION, Entirely Re-written.

FIRST DIVISION. 128 *pages, crown 8vo., with* 100 *engravings, price* 2s.
First Course of Chemical Experiments.
Introductory View of Chemistry. Instructions in Chemical Manipulation. Lessons on the Qualitative Analysis of Salts. Art of Centigrade Testing. Tables of Chemical Equivalents.

SECOND DIVISION. 624 *pages, with* 440 *engravings, price* 10s. 6d.
Chemistry of the Non-Metallic Elements.
Air, Water, the Gases, the Acids, and a Summary of Organic Chemistry; including an extensive Course of CLASS EXPERIMENTS, with Instructions for their successful performance, illustrated by 440 Engravings of the most efficient Apparatus.

Contents of the SECOND DIVISION of CHEMICAL RECREATIONS.

THE RADICAL THEORY:—1. Oxygen. 2. Hydrogen. Compounds with Oxygen. 3. Nitrogen. Compounds with Oxygen and Hydrogen. 4. Carbon. Compounds of Carbon with Oxygen and Hydrogen. Organic Compounds. Salts produced by Organic Radicals. Special examples of Organic Salts. Organic Compounds that contain Nitrogen. Combustion, Fuel, Illumination, Fusion. Gas Furnaces. 5. Sulphur and its Salts. 6. Selenium. 7, 8. Tellurium. 9. Phosphorus. 10, 11. Arsenic. 12, 13. Antimony. 14. Chlorine. 15. Bromine. 16. Iodine. 17. Fluorine. 18. Boron. 19. Silicon. 20, 21. Chromium. 22, 23. Molybdenum. 24, 25. Vanadium. 26. Tungstenum. 27. Titanium. 28. Tantalum. 29. Pelopium. 30. Niobium.

The Work complete, in One Volume, cloth gilt, 12s. 6d.

JOHN J. GRIFFIN & SONS, 22, GARRICK STREET, W.C.

SCIENTIFIC WORKS by JOHN J. GRIFFIN, F.C.S.

CHEMICAL HANDICRAFT:
A CLASSIFIED AND DESCRIPTIVE CATALOGUE OF CHEMICAL APPARATUS,

Suitable for the performance of Class Experiments, for every Process of Chemical Research, and for Chemical Testing in the Arts.

Accompanied by copious Notes, explanatory of the Construction and Use of the Apparatus.

BY JOHN J. GRIFFIN, F.C.S.

In One Large Volume, 8vo., of 472 pp., Illustrated by Sixteen Hundred Engravings on Wood. Price 4s., cloth.

CONTENTS.

Apparatus for Mechanical Operations: Hammers, Mortars, Glass-blowing Apparatus, Supply of Water, &c. Supports for Apparatus, of iron, wood, &c. Weighing and Measuring. Apparatus for determining the Specific Gravity of Liquids: Hydrometers, Alcoholometers, Saccharometers, Urinometers, Specific Gravity Bottles, &c. Pneumatic Apparatus for Chemical Use. Apparatus for the Production and Application of Heat: Portable Furnaces, Spirit Lamps, Blast Spirit Lamps, Blast Oil Lamps, Gas Burners and Gas Furnaces, Gas Combustion Furnaces, Blast Gas Furnaces, Blowing Machines, Gas Blowpipes, Baths for Heating and Drying, Crucibles of all sorts. Tube Operations. Vessels for preparing Solutions: Flasks, Beakers, Jars, Bottles for Chemicals. Filtration, Percolation, Edulcoration. Funnels, Filters, Percolators, Drainers, Pipettes, Syphons, Washing Bottles, &c. Dialysis. Evaporation. Evaporating Basins of all species, Ladles, Cups, &c. Distillation: Retorts, Receivers, Stills, Condensers. The Preparation and Examination of Gases: Gas Bottles and their Fittings, Fitted Bottles for various Gases. Purification of Gases. Pneumatic Troughs, Gas Receivers, Gas Bags, Gas Holders, Condensation and Absorption of Gases, Class Experiments with Gases. Analytical Operations: Application of Chemical Tests, Testing Apparatus, Chemical Tests in solution of graduated strength. Volumetric Analysis: Weights and Measures used. Apparatus: Burettes, Supports for Burettes, Pipettes, Measuring Flasks, Test Mixers, Mixing Jars, Indicators. Special Volumetric Operations. Testing of Carbonates, Chlorides and Iron Salts, Water Test, Assay of Milk, Wine Testing, Sugar-cane Juice Testing, Assay of Zinc Ores, Sets of Apparatus and Test Liquors for Volumetric Analysis. Volumetric Solutions on various standards: the Septem, the Cent. Cube, and the Decem. Urinometry: Sets of Instruments and Tests for the Assay of Urine. Blowpipe Apparatus and Apparatus for Microchemical Operations. Apparatus for Experiments on Coloured Flames. The Spectroscope. Cabinets of Blowpipe Apparatus, both for qualitative and quantitative Analysis. Apparatus for Assaying, and for Metallurgic Operations in general. Marsh's Arsenic Test. Organic Analysis. Collections of Chemical Apparatus arranged for special purposes: some in portable Cabinets. Specimens of Minerals, illustrative of Mineralogy, Geology, and Metallurgy. Crystallography: Collections of Crystal Models. Chemical Books. List of Chemicals. Acids for Exportation. With all other Requisites for Experiments of Demonstration or Research.

JOHN J. GRIFFIN & SONS, 22, GARRICK STREET, W.C.

SCIENTIFIC WORKS by JOHN J. GRIFFIN, F.C.S.

POPULAR GUIDE TO CRYSTALLOGRAPHY.

In demy 8vo., pp. 520, with numerous Figures, price 5s.,

A SYSTEM OF CRYSTALLOGRAPHY, WITH ITS APPLICATION TO MINERALOGY.

By JOHN JOSEPH GRIFFIN, F.C.S.

THIS is the only English work in which the mathematical rules for the examination and description of Crystals are expressed in words at length, as well as in algebraic formulæ; and it contains the only English Catalogue of the Forms and Combinations presented by the discovered Crystals of each species of Mineral.

CONTENTS.

PART I. PRINCIPLES OF CRYSTALLOGRAPHY:—1. Axes of Crystals. 2. Planes of Crystals. 3. Prisms and Pyramids, and their Combinations with one another. 4. Classification of Crystals. 5. Possible limit to the number of Planes that can occur upon Crystals. 6. Crystallographic Notation. 7. Cleavage and Primitive Forms. 8. Forms and Combinations. 9. The Five Zones. 10. Law of Symmetry. 11. Theory of Crystallisation. 12. Use of Spherical Trigonometry in Crystallisation, explained in a popular manner. 13. Inquiry into the variety of Forms and Combinations that occur upon the Crystals of Minerals. Explanation of the Six Systems of Crystallisation:—(1) The Octahedral System; (2) The Pyramidal System; (3) The Rhombohedral System; (4) The Prismatic System; (5) The Oblique Prismatic System; (6) The Doubly-oblique Prismatic System. 14. Brooke's System of Crystallography. 15. Considerations on the utmost possible Abridgment of Crystallographic Notation. 16. Table of Sines and Tangents.

PART II. APPLICATION OF CRYSTALLOGRAPHY TO MINERALOGY:—1. Rose's Tabular Arrangement of known Crystallised Minerals, according to Six Systems of Crystallisation. 2. Catalogue of Crystallised Minerals, showing the Combinations that occur in Nature. 3. Systematic arrangement of the Crystals found in the Mineral Kingdom, with a List of the Minerals common to each Crystal, with an explanation of the Mineralogical Characters employed to discriminate the Minerals that Crystallise in the same Form. 4. Descriptive Catalogue of a Series of 120 Models of Crystals employed to illustrate this system of Crystallography.

MODELS OF CRYSTALS: a Series of 120 Models of Crystals, executed in Biscuit Porcelain, size from 2 to 4 inches in length, most of them representing Crystals that occur among Minerals, and adapted to facilitate the study of Crystallographic Science. *Price of the Set of 120 Models, 42s.*

JOHN J. GRIFFIN & SONS, 22, GARRICK STREET, W.C.

SCIENTIFIC WORKS by JOHN J. GRIFFIN, F.C.S.

THE CHEMICAL TESTING OF WINES AND SPIRITS.

By JOHN JOSEPH GRIFFIN, F.C.S.

In One Volume, Crown 8vo., Illustrated by numerous Woodcuts.

SECOND EDITION, REVISED. Price 5s.

CONTENTS.

Analysis of 41 Wines by the processes described in this work. Table of the weight in grains of the constituents of a gallon of each Wine. Table of the percentages of the same constituents. Determination of the specific gravity of Wines and Spirits.

Alcohol Tables, an entirely new series, founded on the latest analytical investigations. Table of diluted Spirits of from 0 to 12 per cent. of absolute Alcohol by weight, showing, 1, Percentage of Alcohol; 2, Specific Gravity; 3, Weight of a centigallon of the mixture; 4, Weight of absolute Alcohol; 5, Weight of Proof Spirit; 6, Percentage according to Sikes. Table of Diluted Alcohols from 0 to 100 per cent. by weight, with similar details in six columns. Table of Percentages of Alcohol by volume, according to Tralles and Gay-Lussac, compared with percentage of Proof Spirit, according to Sikes. Harmony of these various Alcoholometers. Table for the dilution of Spirits, and for the valuation of Proof Spirits according to Sikes. Series of Problems for calculations respecting Alcohol. Corrections for temperature required by experiments made with Alcohol.

Experimental Determination of the quantity of Alcohol in Wines. Experimental Determination of the quantity of Free Acid in Wines. Investigation of the best means of separating volatile from fixed Acids in Wines. Experimental Determination of the quantity of Sugar in Wines. Process for the separate estimation of Grape Sugar and Cane Sugar. Determination of the amount of Solid Residue left when Wines are evaporated to dryness at 230° Fahr. Determination of the quantity of Ash, or incombustible substances in Wines. Determination of the quantity of Free Alkali contained in the Ash of Wines. Estimate of the neutral Organic bodies contained in Wines. Programme of a Wine Analysis according to the methods described in this work.

Mutual relations of the Constituents of Wines. Conclusions respecting the proportions in which Alcohol, Acid, and Sugar ought to exist with one another to form good Wines. Testing of Spirits. Chemical notes on some special points in the manufacture of Wines. Testing of Must in good seasons and in bad seasons. Correction of Acid Must in bad seasons to render it fit to make good Wine. Preparation of good Wines from unripe Grapes. Winemaking without Grape-juice. Quick process for Maturing Wines. Blending and fortifying of Wines and Spirits. Import duty on Wines.

CHEMICAL APPARATUS and TEST LIQUORS for analysing Wines by the rapid and easy methods described in this work, supplied complete with Balance and Grain Weights. Price £9 9s.

The same collection without Balance and Grain Weights. Price £6 6s.

JOHN J. GRIFFIN & SONS, 22, GARRICK STREET, W.C.

SCIENTIFIC WORKS by JOHN J. GRIFFIN, F.C.S.

CATALOGUES OF SCIENTIFIC APPARATUS.

1. **CABINETS and COLLECTIONS of CHEMICAL** APPARATUS and TESTS, suitable for the Private Study of Elementary Chemistry, for Students in Chemical Schools, for Qualitative Analysis, for the Illustration of Chemical Lectures, for Travelling Mineralogists, Metallurgists, and Engineers, for Toxicology, Urinometry, and other Medical Purposes; for Agricultural Chemists, and other special experimental purposes. Arranged by JOHN J. GRIFFIN, F.C.S. [Extracted from "Chemical Handicraft."] 40 pages, 8vo., sewed, price 6*d*., post free.

2. **APPARATUS for the PRODUCTION and APPLI-**CATION of HEAT by means of Furnaces, Lamps, and Gas-burners, and Fittings suitable for the various Chemical Processes of Evaporation, Drying, Ignition, Fusion, &c. 80 pages, 8vo., illustrated by 300 woodcuts, price 1*s*., post-free.

3. **ILLUSTRATED CATALOGUE of CHEMICAL AP-**PARATUS, suitable for the Private Study of the Science, for Lectures and Class Teaching, and for Analytical Investigations. [A select List from the "Chemical Handicraft."] 64 pages, demy 8vo., with 230 woodcuts, price 1*s*., post-free.

4. **MAGIC LANTERNS,** Apparatus for Dissolving Views, and first-class Sliders, in great variety; with copious Lists of sets of Sliders, and full Instructions for the Use of the Apparatus. 32 pages, 4to., with 14 woodcuts, price 3*d*., post-free.

5. **ACCOUNT of a GAS FURNACE,** suitable for Chemical Fusions at a White Heat, and Assaying Operations by the Muffle without the aid of a Blowing Machine. Price 1*d*., post-free.

6. **LIST OF PURE CHEMICAL TESTS,** and other Chemical Preparations for Experiments of Demonstration or Research. Price 1*d*., post-free.

JOHN J. GRIFFIN & SONS, 22, GARRICK STREET, W.C.

www.ingramcontent.com/pod-product-compliance
Lightning Source LLC
Chambersburg PA
CBHW020910230426
43666CB00008B/1395